THE ANALYSIS OF TIME SERIES:
AN INTRODUCTION

The Analysis of Time Series: An Introduction

C. CHATFIELD

Senior Lecturer in Statistics,
The University of Bath

THIRD EDITION

LONDON   NEW YORK
CHAPMAN AND HALL

First published 1975
Second edition 1980
Third edition 1984
Published by Chapman and Hall Ltd
11 New Fetter Lane, London EC4P 4EE

Published in the USA by
Chapman and Hall
733 Third Avenue, New York NY 10017

© 1975, 1980, 1984 C. Chatfield

Printed in Great Britain by
J. W. Arrowsmith Ltd, Bristol

ISBN 0 412 26030 1

British Library Cataloguing in Publication Data	Library of Congress Cataloging in Publication Data

Chatfield, Christopher
  The analysis of time series.
  1. Time-series analysis
I. Title
519.5'5   QA280
 	
  ISBN 0-412-26030-1
	Chatfield, Christopher.
	  The analysis of time series.
	
	  Bibliography: p.
	  Includes indexes.
	  1. Time-series analysis. I. Title.
	QA280.C4 1984 519.5'5 84–1695
	ISBN 0-412-26030-1 (pbk.)

To
Liz

Alice sighed wearily. 'I think you might do something better with the time,' she said, 'than waste it in asking riddles that have no answers.'

'If you knew Time as well as I do,' said the Hatter, 'you wouldn't talk about wasting *it*. It's *him*.'

'I don't know what you mean,' said Alice.

'Of course you don't!' the Hatter said, tossing his head contemptuously. 'I dare say you never even spoke to Time!'

'Perhaps not,' Alice cautiously replied: 'but I know I have to beat time when I learn music.'

'Ah! that accounts for it,' said the Hatter. 'He won't stand beating.'

Alice's Adventures in Wonderland,
Lewis Carroll.

Contents

CONTENTS

CONTENTS

Preface to Third Edition

The analysis of time series can be a rather difficult topic, and my aim in writing this book has been to provide a comprehensible introduction which considers both theory and practice. Enough theory is given to introduce the concepts of time-series analysis and make the book mathematically interesting. In addition various practical examples are considered so as to help the reader tackle the analysis of real data.

The book can be used as a text for an undergraduate or postgraduate course in time series, or it can be used for self-tuition by research workers. The book assumes a knowledge of basic probability theory and elementary statistical inference. In order to keep the level of mathematics required as low as possible, some mathematical difficulties have been glossed over, particularly in Sections 3.4.8 and 3.5 which the reader is advised to omit at a first reading. Nevertheless a fair level of mathematical sophistication is required in places, as for example in the need to handle Fourier Transforms, although I have helped the reader here by providing a special appendix on this topic.

Although the book is primarily an introductory text, I have nevertheless added appropriate references to further reading and to more advanced topics so that the interested reader can pursue his studies if he wishes. These references are mainly to comprehensible and readily accessible sources rather than to the original attributive references.

One difficulty in writing an introductory textbook is that many practical problems contain at least one feature which is 'non-standard' and these cannot all be envisaged in a book of reasonable length. Thus the reader who has grasped the basic concepts of time-series analysis should always be prepared to use his common sense in tackling a problem. Example 1 of Section 5.5 is a typical situation where common sense has to be applied and also stresses the fact that the first step in any time-series analysis should be to plot the data.

The third edition has much the same structure as the first two editions. The references have been revised throughout and the text has been clarified and updated in many places. However, these changes have been made without altering the Section numbering or the page numbering for the first nine Chapters. Chapter 10 has been revised again, and the major new feature of the third edition is an additional Appendix giving several detailed worked examples. These examples, including candid comments on difficulties, should complement the main text.

I am indebted to many people for assistance in the preparation of this book including V. Barnett, D. R. Cox, K. V. Diprose, R. Fenton, P. R. Fisk, H. Neave, J. Marshall, P. Newbold, M. P. G. Pepper, M. B. Priestley, D. L. Prothero, A. Robinson and C. M. Theobald. Of course, any errors, omissions or obscurities which remain are entirely my responsibility. The author will be glad to hear from any reader who wishes to make constructive comments.

Finally, it is a pleasure to thank Jean Honebon, Doreen Faulds and Elizabeth Aplin for typing the three editions of this book in such an efficient way.

School of Mathematics,   Christopher Chatfield
Bath University,
June 1983.

Abbreviations

AR	Autoregressive
MA	Moving Average
ARMA	Mixed Autoregressive Moving Average
ARIMA	Autoregressive Integrated Moving Average
ac.f	Autocorrelation function
acv.f.	Autocovariance function
FFT	Fast Fourier Transform
$N(\mu,\sigma^2)$	A normally distributed random variable, mean μ, variance σ^2.
χ_ν^2	A chi-squared random variable with ν degrees of freedom

Notation

∇	The difference operator such that $\nabla x_t = x_t - x_{t-1}$.
B	The backward shift operator such that $Bx_t = x_{t-1}$.
E	Expected value or expectation

CHAPTER 1

Introduction

A time series is a collection of observations made sequentially in time. Examples occur in a variety of fields, ranging from economics to engineering, and methods of analysing time series constitute an important area of statistics.

1.1 Examples

We begin with some examples of the sort of time series which arise in practice

(a) Economic time series   Many time series arise in economics, including such series as share prices on successive days, export totals in successive months, average incomes in successive months, company profits in successive years; and so on. Fig. 1.1 shows part of the classic Beveridge wheat price index series which consists of the average wheat price in nearly 50 places in various countries measured in successive years from 1500 to 1869. The complete series is given by Anderson (1971). This series is of particular interest to economic historians, and there is some controversy as to the possible presence of a cycle with period 13.3 years (Granger and Hughes, 1971; Akaike, 1978).

(b) Physical time series   Many types of time series occur in the physical sciences, particularly in meteorology, marine science, and geophysics. Examples are rainfall on successive

1

Figure 1.1   Part of the Beveridge wheat price index series.

days, and air temperature measured in successive hours, days or months. Fig. 1.2 shows the air temperature at Recife, in Brazil. These data are tabulated and analysed in Example 1 of Appendix IV.

Some mechanical recorders take measurements continuously and produce a continuous trace rather than observations at discrete intervals of time. For example in some laboratories it is important to keep temperature and humidity as constant as possible and so devices are installed to measure these variables continuously. Some examples of continuous traces are given in Chapter 7 in Fig. 7.4.

Figure 1.2   Average air temperature at Recife, Brazil, in successive months.

2

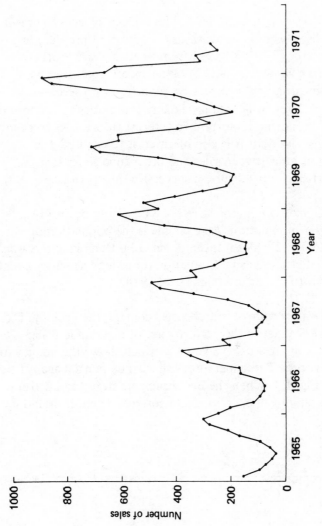

Figure 1.3   Sales of a certain engineering company in successive months.

Continuous recorders often smooth the data to some extent as they are insensitive to high frequency variation.

(c) Marketing time series The analysis of sales figures in successive weeks or months is an important problem in commerce. Fig. 1.3, taken from Chatfield and Prothero (1973a), shows the sales of an engineering product by a certain company in successive months over a seven-year period. Marketing data have many similarities to economic data. It is often important to forecast future sales so as to plan production. It is also of interest to examine the relationship between sales and other time series such as advertising expenditure in successive time periods.

(d) Demographic time series Time series occur in the study of population. An example is the population of England and Wales measured annually. Demographers want to predict changes in population for as long as ten or twenty years into the future (e.g. Brass, 1974).

(e) Process control In process control, the problem is to detect changes in the performance of a manufacturing process by measuring a variable which shows the quality of the process. These measurements can be plotted against time as in Fig. 1.4. When the measurements stray too far from some target value, appropriate corrective action should be

Figure 1.4   A process control chart.

taken to control the process. Special techniques have been
developed for this type of time-series problem and the reader
is referred to a book on statistical quality control (e.g.
Wetherill, 1977).

(f) Binary processes   A special type of time series arises
when observations can take one of only two values, usually
denoted by 0 and 1 (see Fig. 1.5). Time series of this type,
called binary processes, occur particularly in communication
theory. For example the position of a switch, either 'on' or
'off', could be recorded as one or zero respectively.

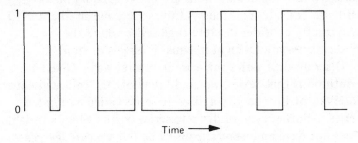

Figure 1.5   A realization of a binary process.

(g) Point processes   A different type of time series occurs
when we consider a series of events occurring 'randomly' in
time. For example we could record the dates of major
railway disasters. A series of events of this type is often called
a *point process* (see Fig. 1.6). For observations of this type,
we are interested in the distribution of the number of events
occurring in a given time-period and also in the distribution
of time intervals between events. Methods of analysing data
of this type are discussed by Cox and Lewis (1966), and will
not be specifically discussed in this book.

*Figure 1.6   A realization of a point process. (✕ denotes
an event.)*

1.2 Terminology

A time series is said to be *continuous* when observations are made continuously in time as in Figs. 1.5 and 7.4. The term continuous is used for series of this type even when the measured variable can only take a discrete set of values, as in Fig. 1.5. A time series is said to be *discrete* when observations are taken only at specific times, usually equally spaced. The term discrete is used for series of this type even when the measured variable is a continuous variable.

In this book we are mainly concerned with discrete time series, where the observations are taken at equal intervals, We will also consider continuous time series, and in Chapter 10 will briefly consider discrete time series where the observations are taken at unequal intervals of time.

Discrete time series can arise in several ways. Given a continuous time series, we could read off the values at equal intervals of time to give a discrete series called a *sampled* series. Another type of discrete series occurs when a variable does not have an instantaneous value but we can *aggregate* (or accumulate) the values over equal intervals of time. Examples of this type are exports measured monthly and rainfall measured daily. Finally some time series are inherently discrete, an example being the dividend paid by a company to shareholders in successive years.

Much statistical theory is concerned with random samples of independent observations. The special feature of time-series analysis is the fact that successive observations are usually *not* independent and that the analysis must take into account the time *order* of the observations. When successive observations are dependent, future values may be predicted from past observations. If a time series can be predicted exactly, it is said to be *deterministic*. But most time series are *stochastic* in that the future is only partly determined by past values. For stochastic series exact predictions are impossible and must be replaced by the idea that future values have a probability distribution which is conditioned by a knowledge of past values.

1.3 Objectives of time-series analysis

There are several possible objectives in analysing a time series. These objectives may be classified as description, explanation, prediction and control, and will be considered in turn.

(a) Description When presented with a time series, the first step in the analysis is usually to plot the data and to obtain simple descriptive measures of the main properties of the series as described in Chapter 2. For example, looking at Fig. 1.3 it can be seen that there is a regular seasonal effect with sales 'high' in winter and 'low' in summer. It also looks as though annual sales are increasing (i.e. show an upward trend). For some series, the variation is dominated by such 'obvious' features, and a fairly simple model, which only attempts to describe trend and seasonal variation, may be perfectly adequate to describe the variation in the time series. For other series, more sophisticated techniques will be required to provide an adequate analysis. Then a more complex model will be constructed, such as the various types of stochastic process described in Chapter 3.

This book devotes a greater amount of space to the more advanced techniques, but this does not mean that elementary descriptive techniques are unimportant. Anyone who tries to analyse a time series, without plotting it first, is asking for trouble. Not only will a graph show up trend and seasonal variation, but it also enables one to look for 'wild' observations or *outliers* which do not appear to be consistent with the rest of the data. The treatment of outliers is a complex subject in which common-sense is as important as theory. The 'outlier' may be a perfectly valid observation in which case the model for the time series will need to take it into account. Alternatively, the outlier may be a freak observation arising, for example, when a recording device goes wrong or when a strike severely affects sales. In the latter case, the outlier needs to be adjusted in some way before further analysis of the data. *Robust* methods (e.g. Kleiner *et al.*, 1979) are designed to be insensitive to outliers.

Another feature to look for in the graph of the time series is the possible presence of turning points, where, for example, an upward trend has suddenly changed to a downward trend. If there is a turning point, different models may have to be fitted to the two parts of the series.

(b) Explanation   When observations are taken on two or more variables, it may be possible to use the variation in one time series to explain the variation in another series. This may lead to a deeper understanding of the mechanism which generated a given time series.

Multiple regression models may be helpful here. In Chapter 9 we also consider the analysis of what are called *linear systems*. A linear system converts an input series to an output series by a linear operation. Given observations on the input and output to a linear system (see Fig. 1.7), one wants to assess the properties of the linear system. For example it is of interest to see how sea level is affected by temperature and pressure, and to see how sales are affected by price and economic conditions.

(c) Prediction   Given an observed time series, one may want to predict the future values of the series. This is an

Figure 1.7   Schematic representation of a linear system.

important task in sales forecasting, and in the analysis of economic and industrial time series. Many writers, including myself, use the terms 'prediction' and 'forecasting' interchangeably, but some authors do not. For example Brown (1963) uses 'prediction' to describe subjective methods and 'forecasting' to describe objective methods, whereas Brass (1974) uses 'forecast' to mean any kind of looking into the future, and 'prediction' to denote a systematic procedure for doing so.

Prediction is closely related to *control* problems in many situations. For example if one can predict that a manufacturing process is going to move off target, then appropriate corrective action can be taken.

d) Control   When a time series is generated which measures the 'quality' of a manufacturing process, the aim of the analysis may be to control the process. Control procedures are of several different kinds. In statistical quality control, the observations are plotted on control charts and the controller takes action as a result of studying the charts. A more sophisticated control strategy has been described by Box and Jenkins (1970). A stochastic model is fitted to the series, future values of the series are predicted, and then the input process variables are adjusted so as to keep the process on target. Many other contributions to control theory have been made by control engineers (see Chapter 10) and the 'controller' may, of course, be some sort of computer.

1.4 Approaches to time-series analysis

This book will describe various approaches to time-series analysis. In Chapter 2, we will describe simple descriptive techniques, which consist of plotting the data, looking for trends, seasonal fluctuations, and so on. Chapter 3 introduces a variety of probability models for time series, while Chapter 4 discusses ways of fitting these models to time series. The major diagnostic tool which is used in Chapter 4 is a function called the *autocorrelation* function which helps to describe the evolution of a process through time. Inference based on this function is often called an analysis in the *time domain*.

Chapter 5 discusses a variety of forecasting procedures. This chapter is not a prerequisite for the rest of the book and the reader may, if he wishes, proceed from Chapter 4 to Chapter 6.

Chapter 6 introduces a function called the *spectral density* function which describes how the variation in a time series may be accounted for by cyclic components at different frequencies. Chapter 7 shows how to estimate this function, a procedure which is called *spectral analysis*. Inference based on the spectral density function is often called an analysis in the *frequency domain*.

Chapter 8 discusses the analysis of two time series, while Chapter 9 extends this work by considering linear systems in which one series is regarded as the input, while the other series is regarded as the output.

Finally Chapter 10 gives a brief introduction and references to a number of other topics, including several recent developments.

1.5 Review of books on time-series

This section gives a brief review of other books on time series. In recent years a number of useful books have been published including an introductory text by Kendall (1976), a clear introduction to spectral analysis (Bloomfield, 1976), a thorough mathematical text at an intermediate level for the univariate case (Anderson, 1971), a book on spectral analysis (Koopmans, 1974), and a text written primarily for engineers (Bendat and Piersol, 1971). The book by Jenkins and Watts (1968) is still a valuable reference source on spectral analysis although it has been somewhat overtaken in parts by the development of the Fast Fourier Transform

algorithm. There are a number of books with a more
theoretical flavour, including Quenouille (1957), Hannan
(1960 and 1970), Brillinger (1975) and Fuller (1976), and
these are more suitable for the advanced reader. The book
by Blackman and Tukey (1959) is of some historical
interest as it was for several years the principal source of
practical information.

The important work of Box and Jenkins (1970) describes
an approach to time-series analysis, forecasting and control
which is based on a particular class of linear stochastic
models. An introduction to their approach is given by
Chapters 3, 4, and 5. Other useful books on forecasting
include Granger and Newbold (1977), Gilchrist (1976), and
Montgomery and Johnson (1977). The subject of control
theory, which is closely related to time series, is briefly
introduced in Chapter 10, but is otherwise outside the scope
of this book. Books on this topic, apart from Box and
Jenkins (1970), include Jacobs (1974), a mathematical
treatise by Whittle (1963) and Wiener's (1949) classic work
which was republished in 1966 with the rather misleading
title of 'Time Series'.

More recent books include a readable introductory text
aimed primarily at social scientists (Gottman, 1981) and a
comprehensive two-volume work by Priestley (1981) which is
particularly strong on spectral analysis, multivariate time
series and non-linear models.

CHAPTER 2

Simple Descriptive Techniques

Statistical techniques for analysing time series, range from the straightforward to the very sophisticated. This chapter describes some fairly simple techniques which will often detect the main properties of a given series. More sophisticated techniques, such as spectral analysis, must await a study of the theory of stationary time series.

2.1 Types of variation

Traditional methods of time-series analysis are mainly concerned with decomposing a series into a trend, a seasonal variation, and other 'irregular' fluctuations. This approach is not always the best approach (Kendall and Stuart, 1966, p. 349) but is still often useful.

These different sources of variation will now be described in more detail.

(a) Seasonal effect   Many time series, such as sales figures and temperature readings, exhibit a variation which is annual in period. This type of variation is easy to understand, and we shall see that it can be measured or removed from the data to give de-seasonalized data.

(b) Other cyclic changes   Apart from seasonal effects,

some time series exhibit variation at a fixed period due to some other physical cause. An example is daily variation in temperature. In addition some time series exhibit oscillations which do not have a fixed period but which are predictable to some extent. For example economic data may be affected by business cycles with a period varying between about 5 and 7 years, although the possible presence of such business cycles is open to some controversy.

(c) Trend This may be loosely defined as 'long term change in the mean'. A difficulty with this definition is deciding what is meant by 'long term'. For example, climatic variables sometimes exhibit cyclic variation over a very long time-period such as 50 years. If one just had 20 years data, this long term oscillation would appear to be a trend, but if several hundred years data were available, the long-term oscillation would be visible. Nevertheless in the short term it may still be more meaningful to think of such a long-term oscillation as a trend. Thus in speaking of a 'trend', we must take into account the number of observations available and make a subjective assessment of what is 'long term'. Granger (1966) defines 'trend in mean' as comprising all frequency components whose wavelength exceeds the length of the observed time series.

(d) Other irregular fluctuations After trend and cyclic variations have been removed from a set of data, we are left with a series of residuals, which may or may not be 'random'. We shall examine various techniques for analysing series of this type to see if some of the apparently irregular variation may be explained in terms of probability models, such as *moving average* or *autoregressive* models which will be introduced in Chapter 3. Alternatively we can see if any cyclic variation is still left in the residuals.

13

2.2 Stationary time-series

A mathematical definition of a *stationary* time series will be given later on. However it is now convenient to introduce the idea of stationarity from an intuitive point of view. Broadly speaking a time series is said to be stationary if there is no systematic change in mean (no trend), if there is no systematic change in variance, and if strictly periodic variations have been removed.

Most of the probability theory of time series is concerned with stationary time series, and for this reason time-series analysis often requires one to turn a non-stationary series into a stationary one so as to use this theory. For example one may remove the trend and seasonal variation from a set of data and then try to model the variation in the residuals by means of a stationary stochastic process.

2.3 Time plot

The first step in analysing a time series is to plot the observations against time. This will show up important features such as trend, seasonality, discontinuities and outliers. Plotting the data is not as easy as it sounds. The choice of scales, the size of the intercept, and the way the points are plotted (e.g. as a continuous line or as separate dots) may substantially affect the way the plot 'looks', and so the analyst must exercise care and judgment.

2.4 Transformations

Plotting the data may indicate if it is desirable to transform them, for example by taking logarithms or square roots. The three main reasons for making a transformation are

(a) To stabilize the variance   If there is trend in the series and the variance appears to increase with the mean then it may be advisable to transform the data. In particular if the

14

standard deviation is directly proportional to the mean, a logarithmic transformation is indicated.

(b) To make the seasonal effect additive   If there is a trend in the series and the size of the seasonal effect appears to increase with the mean then it may be advisable to transform the data so as to make the seasonal effect constant. In particular if the size of the seasonal effect is directly proportional to the mean, then the seasonal effect is said to be multiplicative and a logarithmic transformation is appropriate to make the effect additive. However this transformation will only stabilize the variance if the error term is *also* thought to be multiplicative, a point which is sometimes overlooked.

Three seasonal models in common use are:

A) $x_t = m_t + s_t + \epsilon_t$

B) $x_t = m_t s_t + \epsilon_t$

C) $x_t = m_t s_t \epsilon_t$

where x_t = observation at time t
m_t = current mean
s_t = seasonal effect
ϵ_t = random error.

Model A is completely additive and no transformation is required. Model C is completely multiplicative and a logarithmic transformation is certainly appropriate. Model B has multiplicative seasonality and additive error and the relative size of these effects determines whether a transformation is desirable. Further difficulties arise in practice if the seasonal and/or error terms are neither multiplicative nor additive. For example the seasonal effect may increase with the mean but not at such a fast rate so that it is somewhere in-between being multiplicative and additive. A mixed additive–multiplicative seasonal model is introduced by Durbin and Murphy (1975).

15

(c) To make the data normally distributed This is discussed by Granger and Newbold (1977, Section 9.1).

Further remarks on transformations are made in Example 3 of Appendix IV.

2.5 Analysing series which contain a trend

The analysis of a time series which exhibits 'long term' change in mean depends on whether one wants to (a) measure the trend and/or (b) remove the trend in order to analyse local fluctuations. With seasonal data, it is a good idea to start by calculating successive yearly averages as these will provide a simple description of the underlying trend. An approach of this type is sometimes perfectly adequate, particularly if the trend is fairly small, but sometimes a more sophisticated approach is desired and then the following techniques can be considered.

2.5.1 Curve-fitting A traditional method of dealing with non-seasonal data which contains a trend, particularly yearly data, is to fit a simple function such as a polynomial curve (linear, quadratic, etc), a Gompertz curve or a logistic curve (e.g. see Gregg *et al.*, 1964; Harrison and Pearce, 1972; Button, 1974; Levenbach and Reuter, 1976). The Gompertz curve is given by

$$\log x_t = a - br^t$$

where a, b, r are parameters with $0 < r < 1$, while the logistic curve is given by

$$x_t = a/[1 + b \exp(-ct)].$$

Both these curves are S-shaped and approach an asymptotic value as $t \to \infty$. Fitting the curves to data may lead to non-linear simultaneous equations.

For all curves of this type, the fitted function provides a measure of the trend, and the residuals provide an estimate of local fluctuations, where the residuals are the differences between the observations and the corresponding values of the fitted curve.

2.5.2 Filtering

A second procedure for dealing with a trend is to use a *linear filter* which converts one time series, $\{x_t\}$, into another, $\{y_t\}$, by the linear operation

$$y_t = \sum_{r=-q}^{+s} a_r x_{t+r}$$

where $\{a_r\}$ are a set of weights. In order to smooth out local fluctuations and estimate the local mean, we should clearly choose the weights so that $\Sigma a_r = 1$, and then the operation is often referred to as a *moving average*. Moving averages are discussed in detail by Kendall (1976, Chapters 3 and 4), and we will only provide a brief introduction. Moving averages are often symmetric with $s = q$ and $a_j = a_{-j}$. The simplest example of a symmetric smoothing filter is the simple moving average for which $a_r = 1/(2q + 1)$ for $r = -q, \ldots, +q$, and $\mathrm{Sm}(x_t) =$ smoothed value of x_t

$$= \frac{1}{2q.+ 1} \sum_{r=-q}^{+q} x_{t+r}$$

The simple moving average is not generally recommended by itself for measuring trend, although it can be useful for removing seasonal variation. Another example is provided by the case where the $\{a_r\}$ are successive terms in the expansion of $(\frac{1}{2} + \frac{1}{2})^{2q}$. Thus when $q = 1$, the weights are $a_{-1} = a_{+1} = \frac{1}{4}, a_0 = \frac{1}{2}$. As q gets large, the weights approximate to a normal curve.

A third example is Spencer's 15 point moving average, which is used for smoothing mortality statistics to get life-tables (Tetley, 1946). This covers 15 consecutive points with $q = 7$ and the symmetric weights are

$$\frac{1}{320} [-3, -6, -5, 3, 21, 46, 67, 74, \ldots].$$

A fourth example, called the Henderson moving average, is described by Kenny and Durbin (1982).

Another general approach is to fit a polynomial curve, not to the whole series, but to fit polynomials of the same order to different parts. For example one could fit a cubic polynomial to the first $(2q + 1)$ data points, where $q \geqslant 2$, use this polynomial to determine the interpolated value at the middle of the range (i.e. $t = q + 1$), then repeat the operation using the $(2q + 1)$ data points from $t = 2$ to $t = (2q + 2)$ and so on. An alternative class of piece-wise polynomials, is the class of *spline* functions (see Wold, 1974).

Suppose we have N observations, $x_1, \ldots, x_N$. For a symmetric filter, we can calculate $\mathrm{Sm}(x_t)$ for $t = q + 1$ up to $t = N - q$, but as a consequence lose $2q$ pieces of data. In some situations this is not important but in other situations it is particularly important to get smoothed values up to $t = N$. Kendall and Stuart (1966) discuss the end-effects problem. One can either project the smoothed values by eye, fit a polynomial to the last few smoothed values, or alternatively use an asymmetric filter such as

$$\mathrm{Sm}(x_t) = \sum_{j=0}^{q} a_j x_{t-j}.$$

The popular technique known as *exponential smoothing* (see also Section 5.2.2) takes $q = \infty$, and

$$a_j = \alpha(1 - \alpha)^j$$

where α is a constant such that $0 < \alpha < 1$.

Having estimated the trend, we can look at the local fluctuations by examining

$$\mathrm{Res}(x_t) = \text{residual from smoothed value}$$

$$= x_t - \mathrm{Sm}(x_t)$$

$$= \sum_{r=-q}^{+s} b_r x_{t+r}.$$

18

This is also a linear filter, and if

$$Sm(x_t) = \sum_{r=-q}^{+s} a_r x_{t+r}.$$

with $\Sigma a_r = 1$, then $\Sigma b_r = 0$, $b_0 = 1 - a_0$, and $b_r = -a_r$ for $r \neq 0$.

How do we choose the appropriate filter? The answer to this question really requires considerable experience plus a knowledge of the frequency aspects of time-series analysis which will be discussed in later chapters. As the name implies, filters are usually designed to produce an output with emphasis on variation at particular frequencies. For example to get smoothed values we want to remove the local fluctuations which constitute what is called the high-frequency variation. In other words we want what is called a *low-pass* filter. To get $Res(x_t)$, we want to remove the long-term fluctuations or the low-frequency variation. In other words we want what is called a *high-pass* filter. The Slutsky (or Slutsky-Yule) effect is related to this problem. Slutsky showed that by operating on a completely random series with both averaging and differencing procedures one could induce sinusoidal variation in the data, and he went on to suggest that apparently periodic behaviour in some economic time series might be accounted for by the smoothing procedures used to form the data. We will return to this question later.

FILTERS IN SERIES   Very often a smoothing procedure is carried out in two or more stages — so that one has in effect several linear filters in series. For example two filters in series may be represented as follows:

Filter I with weights $\{a_{j1}\}$ acts on $\{x_t\}$ to produce $\{y_t\}$.
Filter II with weights $\{a_{j2}\}$ acts on $\{y_t\}$ to produce $\{z_t\}$.

Now $z_t = \sum_j a_{j2} y_{t+j}$

$$= \sum_j a_{j2} \sum_r a_{r1} x_{t+j+r}$$

$$= \sum c_j x_{t+j}$$

where

$$c_j = \sum_r a_{r1} a_{(j-r)2}$$

are the weights for the overall filter. The weights $\{c_j\}$ are obtained by a procedure called *convolution* and we write

$$\{c_j\} = \{a_{r1}\} \quad * \quad \{a_{r2}\}$$

where * represents convolution.

For example the filter ($\frac{1}{4}$, $\frac{1}{2}$, $\frac{1}{4}$) may be written as

$$(\tfrac{1}{4}, \tfrac{1}{2}, \tfrac{1}{4}) = (\tfrac{1}{2}, \tfrac{1}{2}) * (\tfrac{1}{2}, \tfrac{1}{2}).$$

Given a series $x_1, \ldots, x_N$, this smoothing procedure is best done in three stages by adding successive pairs of observations twice and then dividing by 4, as follows:

Observations	Stage I	Stage II	Stage III
x_1			
	$x_1 + x_2$		
x_2		$x_1 + 2x_2 + x_3$	$(x_1 + 2x_2 + x_3)/$
	$x_2 + x_3$		.
x_3		$x_2 + 2x_3 + x_4$	.
	$x_3 + x_4$		.
x_4		$x_3 + 2x_4 + x_5$	.
	$x_4 + x_5$	.	.
x_5		.	.
	$x_5 + x_6$	.	.
x_6	.	.	
.	.	.	

The Spencer 15 point moving average is actually a convolution of 4 filters, namely

$$(\tfrac{1}{4}, \tfrac{1}{4}, \tfrac{1}{4}, \tfrac{1}{4}) * (\tfrac{1}{4}, \tfrac{1}{4}, \tfrac{1}{4}, \tfrac{1}{4}) * (\tfrac{1}{5}, \tfrac{1}{5}, \tfrac{1}{5}, \tfrac{1}{5}, \tfrac{1}{5}) * (-\tfrac{3}{4}, \tfrac{3}{4}, 1, \tfrac{3}{4}, -\tfrac{3}{4}).$$

2.5.3 Differencing   A special type of filtering, which is particularly useful for removing a trend, is simply to difference a given time series until it becomes stationary. This method is particularly stressed by Box and Jenkins (1970). For non-seasonal data, first-order differencing is usually sufficient to attain apparent stationarity, so that the new series $\{y_1, \ldots, y_{N-1}\}$ is formed from the original series $\{x_1, \ldots, x_N\}$ by

$$y_t = x_{t+1} - x_t = \nabla x_{t+1}$$

First-order differencing is now widely used in economics. Occasionally second-order differencing is required using the operator ∇^2 where

$$\nabla^2 x_{t+2} = \nabla x_{t+2} - \nabla x_{t+1} = x_{t+2} - 2x_{t+1} + x_t.$$

2.6 Seasonal fluctuations

The analysis of time series which exhibit seasonal fluctuations depends on whether one wants to

  (a) measure the seasonals

and/or

  (b) eliminate them.

For series showing little trend, it is usually adequate to simply calculate the average for each month (or quarter, or 4-week period etc.) and compare it with the overall average figure, either as a difference or as a ratio.

For series which do contain a substantial trend, a more sophisticated approach may be required. With monthly data, the commonest way of eliminating the seasonal effect is to calculate

$$Sm(x_t) = \frac{\tfrac{1}{2}x_{t-6} + x_{t-5} + x_{t-4} + \ldots + x_{t+5} + \tfrac{1}{2}x_{t+6}}{12}.$$

Note that the sum of the coefficients is 1. A simple moving average cannot be used as this would span 12 months and would not be centred on an integer value of t. A simple moving average over 13 months cannot be used, as this would give twice as much weight to the month appearing at both ends. Durbin (1963) discusses this filter and simple ways of doing the computation. For quarterly data, the seasonal effect can be eliminated by calculating

$$\text{Sm}(x_t) = \frac{\frac{1}{2}x_{t-2} + x_{t-1} + x_t + x_{t+1} + \frac{1}{2}x_{t+2}}{4}.$$

For 4-weekly data, one *can* use a simple moving average over 13 successive observations.

The seasonal effect can be estimated by $x_t - \text{Sm}(x_t)$ or $x_t/\text{Sm}(x_t)$ depending on whether the seasonal effect is thought to be additive or multiplicative. If the seasonal variation stays roughly the same size regardless of the mean level, then it is said to be additive, but if it increases in size in direct proportion to the mean level, then it is said to be multiplicative. The time plot should be examined to see which description is the better.

A check should be made that the seasonals are reasonably stable, and then the average monthly (or quarterly, etc.) effects can be calculated. If necessary a transformation should be made — as previously discussed.

A seasonal effect can also be eliminated by differencing (see Sections 4.6, 5.2.4; Box and Jenkins, 1970). For example with monthly data one can employ the operator ∇_{12} where

$$\nabla_{12}\, x_t = x_t - x_{t-12}.$$

Alternative methods of seasonal adjustment are reviewed by Pierce (1980). These include the widely used X-11 method (Cleveland and Tiao, 1976; Shiskin and Plewes, 1978). The possible presence of calendar effects should also be considered (Cleveland and Durbin, 1982).

2.7 Autocorrelation

An important guide to the properties of a time series is provided by a series of quantities called sample autocorrelation coefficients, which measure the correlation between observations at different distances apart. These coefficients often provide insight into the probability model which generated the data. We assume that the reader is familiar with the ordinary correlation coefficient, namely that given N pairs of observations on two variables x and y, the correlation coefficient is given by

$$r = \frac{\Sigma(x_i - \bar{x})(y_i - \bar{y})}{\sqrt{[\Sigma(x_i - \bar{x})^2 \Sigma(y_i - \bar{y})^2]}} \tag{2.1}$$

A similar idea can applied to time series to see if successive observations are correlated.

Given N observations $x_1, \ldots, x_N$, on a discrete time series we can form $(N - 1)$ pairs of observations, namely (x_1, x_2), $(x_2, x_3), \ldots, (x_{N-1}, x_N)$. Regarding the first observation in each pair as one variable, and the second observation as a second variable, the correlation coefficient between x_t and x_{t+1} is given by

$$r_1 = \frac{\sum\limits_{t=1}^{N-1} (x_t - \bar{x}_{(1)})(x_{t+1} - \bar{x}_{(2)})}{\sqrt{[\sum\limits_{t=1}^{N-1} (x_t - \bar{x}_{(1)})^2 \sum\limits_{t=1}^{N-1} (x_{t+1} - \bar{x}_{(2)})^2]}} \tag{2.2}$$

by analogy with equation (2.1), where

$$\bar{x}_{(1)} = \sum\limits_{t=1}^{N-1} x_t/(N - 1)$$

is the mean of the first $(N - 1)$ observations and

$$\bar{x}_{(2)} = \sum\limits_{t=2}^{N} x_t/(N - 1)$$

is the mean of the last $(N - 1)$ observations.

As the coefficient given by equation (2.2) measures correlation between successive observations it is called an *autocorrelation* coefficient or serial correlation coefficient.

Equation (2.2) is rather complicated, and so as $\bar{x}_{(1)} \simeq \bar{x}_{(2)}$, it is usually approximated by

$$r_1 = \frac{\sum\limits_{t=1}^{N-1} (x_t - \bar{x})(x_{t+1} - \bar{x})}{(N-1) \sum\limits_{t=1}^{N} (x_t - \bar{x})^2 / N} \tag{2.3}$$

where $\bar{x} = \sum\limits_{t=1}^{N} x_t / N$

= overall mean.

Some authors also drop the factor $N/(N-1)$ which is close to one for large N to give the even simpler formula

$$r_1 = \frac{\sum\limits_{t=1}^{N-1} (x_t - \bar{x})(x_{t+1} - \bar{x})}{\sum\limits_{t=1}^{N} (x_t - \bar{x})^2} \tag{2.4}$$

and this is the form that will be used in this book.

In a similar way we can find the correlation between observations a distance k apart, which is given by

$$r_k = \frac{\sum\limits_{t=1}^{N-k} (x_t - \bar{x})(x_{t+k} - \bar{x})}{\sum\limits_{t=1}^{N} (x_t - \bar{x})^2} \tag{2.5}$$

This is called the autocorrelation coefficient at lag k.

In practice the autocorrelation coefficients are usually calculated by computing the series of autocovariance coefficients, $\{c_k\}$, which we define by analogy with the usual covariance formula as

$$c_k = \frac{1}{N} \sum_{t=1}^{N-k} (x_t - \bar{x})(x_{t+k} - \bar{x}) \qquad (2.6)$$

= autocovariance coefficient at lag k.

We then compute

$$r_k = c_k/c_0 \qquad (2.7)$$

for $k = 1, 2, \ldots, m$, where $m < N$. There is usually little point in calculating r_k for values of k greater than about $N/4$.

Note that some authors suggest

$$c_k = \frac{1}{N-k} \sum_{t=1}^{N-k} (x_t - \bar{x})(x_{t+k} - \bar{x})$$

rather than equation 2.6, but there is little difference for large N (see Section 4.1).

2.7.1 The correlogram   A useful aid in interpreting a set of autocorrelation coefficients is a graph called a correlogram in which r_k is plotted against the lag k. Examples are given in Figs. 2.1–2.3. A visual inspection of the correlogram is often very helpful.

2.7.2 Interpreting the correlogram   Interpreting the meaning of a set of autocorrelation coefficients is not always easy. Here we offer some general advice.

(a) A random series   If a time series is completely random, then for large N, $r_k \cong 0$ for all non-zero values of k. In fact we will see later that for a random time series r_k is approximately $N(0, 1/N)$, so that, if a time series is random, 19 out of 20 of the values of r_k can be expected to lie between $\pm 2/\sqrt{N}$. However if one plots say the first 20 values of r_k then one can expect to find one 'significant' value on average even when the time series really is random. This

spotlights one of the difficulties in interpreting the correlogram, in that a large number of coefficients is quite likely to contain one (or more) 'unusual' results, even when no real effects are present. (See also Section 4.1.)

(b) Short-term correlation Stationary series often exhibit short-term correlation characterized by a fairly large value of r_1 followed by 2 or 3 more coefficients which, while significantly greater than zero, tend to get successively smaller. Values of r_k for longer lags tend to be approximately zero. An example of such a correlogram is shown in Fig. 2.1. A time series which gives rise to such a correlogram, is one

Figure 2.1   A time series showing short-term correlation together with its correlogram

for which an observation above the mean tends to be followed by one or more further observations above the mean, and similarly for observations below the mean. A model called an autoregressive model, may be appropriate for series of this type.

(c) Alternating series   If a time series has a tendency to alternate, with successive observations on different sides of the overall mean, then the correlogram also tends to alternate. The value of r_1 will be negative. However the value of r_2 will be positive as observations at lag 2 will tend to be on the same side of the mean. A typical alternating time series together with its correlogram is shown in Fig. 2.2.

(d) Non-stationary series   If a time series contains a trend, then the values of r_k will not come down to zero except for very large values of the lag. This is because an observation on

Figure 2.2   An alternating time series together with its correlogram.

Figure 2.3   A non-stationary time series together with its correlogram.

one side of the overall mean tends to be followed by a large number of further observations on the same side of the mean because of the trend. A typical non-stationary time series together with its correlogram is shown in Fig. 2.3. Little can be inferred from a correlogram of this type as the trend dominates all other features. In fact the sample autocorrelation function, $\{r_k\}$, should only be calculated for *stationary* time series (see Chapters 3 and 4) and so any trend should be removed before calculating $\{r_k\}$.

(e) Seasonal fluctuations   If a time series contains a seasonal fluctuation, then the correlogram will also exhibit an oscillation at the same frequency. For example with monthly observations, r_6 will be 'large' and negative, while r_{12} will be 'large' and positive. In particular if x_t follows a sinusoidal

28

pattern, then so does r_k. For example, if

$$x_t = a \cos t\omega$$

where a is a constant and the frequency ω is such that $0 < \omega < \pi$, then it can easily be shown that (see exercise 3)

$$r_k \cong \cos k\omega \text{ for large } N.$$

Fig. 2.4(a) shows the correlogram of the monthly air temperature data shown in Fig. 1.2

Figure 2.4   The correlogram of monthly observations on air temperature at Recife (a) for the raw data, (b) for the seasonally adjusted data.

The sinusoidal pattern of the correlogram is clearly evident, but for seasonal data of this type the correlogram provides little extra information as the seasonal pattern is clearly evident in the time plot of the data.

If the seasonal variation is removed from seasonal data, then the correlogram may provide useful information. The seasonal variation was removed from the air temperature data by the simple procedure of calculating the 12 monthly averages and subtracting the appropriate one from each individual observation. The correlogram of the resulting series (Fig. 2.4(b)) shows that the first three coefficients are significantly different from zero. This indicates short-term correlation in that a month which is colder than the average for that month will tend to be followed by one or two further months which are colder than average.

(f) Outliers   If a time series contains one or more outliers, the correlogram may be seriously affected and it is usually advisable to adjust outliers as described in Section 1.3. For example, if there is one outlier in the time series and it is not adjusted, then the plot of x_t against x_{t+k} will contain *two* 'extreme' points which will tend to depress the sample correlation coefficients towards zero. If there are two outliers, this effect is even more noticeable except when the lag equals the distance between the outliers when a spuriously large correlation may occur.

(g) General remarks   Clearly considerable experience is required in interpreting autocorrelation coefficients. In addition we need to study the probability theory of stationary series and discuss the classes of model which may be appropriate. We must also discuss the sampling properties of r_k. These topics will be covered in the next two chapters and we shall then be in a better position to interpret the correlogram of a given time series.

2.8 Other tests of randomness

In most cases, a visual examination of the graph of a time series is enough to see that it is *not* random. However it is occasionally desirable to test a stationary time series for 'randomness'. In other words one wants to test if $x_1, \ldots, x_N$ could have arisen in that order by chance by taking a simple random sample size N from a population with unknown characteristics. A variety of tests exist for this purpose and they are described by Kendall and Stuart (1966, Section 45.15) and by Kendall (1976; Chapter 2). For example one can examine the number of times there is a local maximum or minimum in the time series. A local maximum is defined to be any observation x_t such that $x_t > x_{t-1}$ and $x_t > x_{t+1}$. A converse definition applies to local minima. If the series really is random one can work out the expected number of turning points and compare it with the observed value. Tests of this type will not be described here. I have always found it convenient to simply examine the correlogram (and possibly the spectral density function) of a given time series to see if it is random.

Having fitted a model to a non-random series, one often wants to see if the residuals are random. Testing residuals for randomness is a somewhat different problem and will be discussed in Section 4.7.

Exercises

1. The following data show the sales of company X in successive 4-week periods over 1967–1970.

	I	II	III	IV	V	VI	VII	VIII	IX	X	XI	XII	XIII
1967	153	189	221	215	302	223	201	173	121	106	86	87	108
1968	133	177	241	228	283	255	238	164	128	108	87	74	95
1969	145	200	187	201	292	220	233	172	119	81	65	76	74
1970	111	170	243	178	248	202	163	139	120	96	95	53	94

    (a) Plot the data.

    (b) Assess the trend and seasonal effects.

2. Sixteen successive observations on a stationary time series are as follows:-

   1.6, 0.8, 1.2, 0.5, 0.9, 1.1, 1.1, 0.6, 1.5, 0.8, 0.9, 1.2, 0.5, 1.3, 0.8, 1.2

   (a) Plot the observations.

   (b) Looking at the graph, guess an approximate value for the autocorrelation coefficient at lag 1.

   (c) Plot x_t against x_{t+1}, and again try to guess the value of r_1.

   (d) Calculate r_1.

3. If $x_t = a \cos t\omega$ where a is a constant and ω is a constant in $(0, \pi)$, show that $r_k \to \cos k\omega$ as $N \to \infty$.

4. The first ten sample autocorrelation coefficients of 400 'random' numbers are $r_1 = 0.02$, $r_2 = 0.05$, $r_3 = -0.09$, $r_4 = 0.08$, $r_5 = -0.02$, $r_6 = 0.00$, $r_7 = 0.12$, $r_8 = 0.06$, $r_9 = 0.02$, $r_{10} = -0.08$. Is there any evidence of non-randomness?

CHAPTER 3

Probability Models for Time Series

3.1 Stochastic processes

This chapter describes a variety of probability models for time series, which are collectively called stochastic processes. Most physical processes in the real world involve a random or stochastic element in their structure, and a stochastic process can be described as 'a statistical phenomenon that evolves in time according to probabilistic laws'. Well-known examples are the length of a queue, the size of a bacterial colony, and the air temperature on successive days at a particular site. Many authors use the term 'stochastic process' to describe both the real physical process and a mathematical model of it. The word 'stochastic', which is of Greek origin, is used to mean 'pertaining to chance', and many writers use 'random process' as a synonym for stochastic process.

Mathematically, a stochastic process may be a defined as a collection of random variables $\{X(t), t \in T\}$, where T denotes the set of time-points at which the process is defined. We will denote the random variable at time t by $X(t)$ if T is continuous (usually $-\infty < t < \infty$), and by X_t if T is discrete (usually $t = 0, \pm 1, \pm 2, \ldots$). Thus a stochastic process is a collection of random variables which are ordered in time. For a single outcome of the process we have only one observation on each random variable and these values evolve in time according to probabilistic laws.

THE ANALYSIS OF TIME SERIES

The theory of stochastic processes has been extensively developed and is discussed in many books including Papoulis (1965), written primarily for engineers, Parzen (1962), Cox and Miller (1968, especially Chapter 7), Yaglom (1962) and Grimmett and Stirzaker (1982). In this chapter we concentrate on those aspects particularly relevant to time-series analysis.

In most statistical problems we are concerned with estimating the properties of a population from a sample. In time-series analysis, however, it is often impossible to make more than one observation at a given time so that we only have one observation on the random variable at time t. Nevertheless we may regard the observed time series as just one example of the infinite set of time series which might have been observed. This infinite set of time series is sometimes called the *ensemble*. Every member of the ensemble is a possible *realization* of the stochastic process. The observed time series can be thought of as one particular realization, and will be denoted by $x(t)$ for ($0 \leqslant t \leqslant T$) if observations are continuous, and by x_t for $t = 1, \ldots, N$, if observations are discrete.

Because there is only a notional population, time-series analysis is essentially concerned with evaluating the properties of the probability model which generated the observed time series.

One way of describing a stochastic process is to specify the joint probability distribution of $X_{t_1}, \ldots, X_{t_n}$ for any set of times $t_1, \ldots, t_n$ and any value of n. But this is rather complicated and is not usually attempted in practice. A simpler, more useful way of describing a stochastic process is to give the *moments* of the process, particularly the first and second moments, which are called the mean, variance and autocovariance functions. These will now be defined for discrete time, with similar definitions applying in continuous time.

34

Mean   The mean function, $\mu(t)$, is defined by

$$\mu(t) = E(X_t).$$

Variance   The variance function, $\sigma^2(t)$, is defined by

$$\sigma^2(t) = \text{Var}(X_t).$$

Autocovariance   The autocovariance function, $\gamma(t_1, t_2)$, is defined by

$$\gamma(t_1, t_2) = \text{Cov}(X_{t_1}, X_{t_2})$$
$$= E\{[X_{t_1} - \mu(t_1)][X_{t_2} - \mu(t_2)]\}$$

The variance function is a special case of $\gamma(t_1, t_2)$ when $t_1 = t_2$. (Readers who are unfamiliar with 'covariance' should read Appendix III.)

Higher moments of a stochastic process may be defined in an obvious way, but are rarely used in practice, since a knowledge of the two functions $\mu(t)$ and $\gamma(t_1, t_2)$ is often adequate.

3.2 Stationary processes

An important class of stochastic processes are those which are *stationary*. A heuristic idea of stationarity was introduced in Section 2.2.

A time series is said to be *strictly stationary* if the joint distribution of $X(t_1), \ldots, X(t_n)$ is the same as the joint distribution of $X(t_1 + \tau), \ldots, X(t_n + \tau)$ for all $t_1, \ldots, t_n, \tau$. In other words shifting the time origin by an amount τ has no effect on the joint distributions which must therefore depend only on the intervals between $t_1, t_2, \ldots, t_n$. The above definition holds for any value of n. In particular, if $n = 1$, it implies that the distribution of $X(t)$ must be the same for all t so that

$$\mu(t) = \mu$$
$$\sigma^2(t) = \sigma^2$$

are both constants which do not depend on the value of t.

Furthermore if $n = 2$, the joint distribution of $X(t_1)$ and $X(t_2)$ depends only on $(t_2 - t_1)$ which is called the *lag*. Thus the autocovariance function, $\gamma(t_1, t_2)$, also depends only on $(t_2 - t_1)$ and may be written as $\gamma(\tau)$ where

$$\gamma(\tau) = E[(X(t) - \mu)(X(t + \tau) - \mu)]$$

is called the autocovariance coefficient at lag τ. In future 'autocovariance function' will be abbreviated to acv.f.

The size of the autocovariance coefficients depend on the units in which $X(t)$ is measured. Thus, for interpretative purposes, it is useful to standardize the acv.f. to produce a function called the *autocorrelation* function which is given by

$$\rho(\tau) = \gamma(\tau)/\gamma(0),$$

and which measures the autocorrelation between $X(t)$ and $X(t + \tau)$. Its empirical counterpart was introduced in Section 2.7. In future, 'autocorrelation function' will be abbreviated to ac.f. Note that the argument, τ, of $\gamma(\tau)$ and $\rho(\tau)$ is discrete if the time series is discrete and continuous if the time series is continuous.

At first sight it may seem surprising to suggest that there are processes for which the distribution of $X(t)$ should be the same for all t. However readers with some knowledge of stochastic processes will know that there are many processes $\{X(t)\}$ which have what is called an *equilibrium* distribution as $t \to \infty$, in which the probability distribution of $X(t)$ tends to a limit which does *not* depend on the initial conditions. Thus once such a process has been running for some time, the distribution of $X(t)$ will change very little. Indeed if the initial conditions are specified to be identical to the equilibrium distribution, the process is stationary in time and the equilibrium distribution is then the stationary distribution of the process. Of course the *conditional* distribution of $X(t_2)$ given that $X(t_1)$ has taken a particular value, say $x(t_1)$, may be quite different from the stationary distribution, but this is perfectly consistent with the process being stationary.

3.2.1 Second-order stationarity

In practice it is often useful to define stationarity in a less restricted way than that described above. A process is called second-order stationary (or weakly stationary) if its mean is constant and its acv.f. depends only on the lag, so that

$$E[X(t)] = \mu$$

and

$$\text{Cov}[X(t), X(t + \tau)] = \gamma(\tau).$$

No assumptions are made about higher moments than those of second order. Note that the above definition implies that the variance, as well as the mean, is constant. Also note that both the variance and the mean must be finite.

This weaker definition of stationarity will be used from now on, as many of the properties of stationary processes depend only on the structure of the process as specified by its first and second moments (Bartlett, 1966; Chapter 6). One important class of processes where this is particularly true is the class of *normal* processes where the joint distribution of $X(t_1), \ldots, X(t_n)$ is multivariate normal for all $t_1, \ldots, t_n$. The multivariate normal distribution is completely characterized by its first and second moments, and hence by $\mu(t)$ and $\gamma(t_1, t_2)$, and so it follows that second-order stationarity implies strict stationarity for normal processes. However for processes which are very 'non-normal', μ and $\gamma(\tau)$ may not adequately describe the process.

3.3 The autocorrelation function

We have already noted in Section 2.7 that the sample autocorrelation coefficients of an observed time series are an important set of statistics for describing the time series. Similarly the (theoretical) autocorrelation function (ac.f.) of a stationary stochastic process is an important tool for assessing its properties. This section investigates the general properties of the ac.f.

Suppose a stationary stochastic process $X(t)$ has mean μ, variance σ^2, acv.f. $\gamma(\tau)$, and ac.f. $\rho(\tau)$. Then

$$\rho(\tau) = \gamma(\tau)/\gamma(0) = \gamma(\tau)/\sigma^2.$$

Note that $\rho(0) = 1$.

Property 1   The ac.f. is an *even* function of the lag in that

$$\rho(\tau) = \rho(-\tau).$$

This is easily demonstrated since $\rho(\tau) = \gamma(\tau)/\sigma^2$ and $X(t)$ is stationary so that

$$\gamma(\tau) = \text{Cov}[X(t), X(t + \tau)]$$
$$= \text{Cov}[X(t - \tau), X(t)]$$
$$= \gamma(-\tau).$$

Property 2 $|\rho(\tau)| \leqslant 1$. This can be proved by noting that

$$\text{Var}[\lambda_1 X(t) + \lambda_2 X(t + \tau)] \geqslant 0$$

for any constants λ_1, λ_2. This variance is equal to

$$\lambda_1^2 \, \text{Var}[X(t)] + \lambda_2^2 \, \text{Var}[X(t + \tau)] +$$
$$2\lambda_1 \lambda_2 \, \text{Cov}[X(t), X(t + \tau)] = (\lambda_1^2 + \lambda_2^2)\sigma^2 + 2\lambda_1 \lambda_2 \gamma(\tau)$$

When $\lambda_1 = \lambda_2 = 1$, we find

$$\gamma(\tau) \geqslant -\sigma^2$$

so that $\rho(\tau) \geqslant -1$.
    When $\lambda_1 = 1, \lambda_2 = -1$, we find

$$\sigma^2 \geqslant \gamma(\tau)$$

so that $\rho(\tau) \leqslant +1$.

Property 3   Non-uniqueness. For a particular acv.f. there corresponds only one possible stationary normal process, as we have already remarked in Section 3.2.1 that a stationary normal process is completely determined by its mean, variance and ac.f. However, it is always possible to find many non-normal processes with the same ac.f. and this creates a further difficulty in interpreting sample ac.f's. Jenkins and Watts (1968; p. 170) give an example of two different stochastic processes which have the same ac.f.

3.4 Some useful stochastic processes

This section describes several different types of stochastic process which are sometimes useful in setting up a model for a time series.

3.4.1 A purely random process

A discrete process $\{Z_t\}$ is called a purely random process if the random variables $\{Z_t\}$ are a sequence of mutually independent, identically distributed variables. From the definition it follows that the process has constant mean and variance and that

$$\begin{aligned}\gamma(k) &= \mathrm{Cov}(Z_t, Z_{t+k}) \\ &= 0 \quad \text{for} \quad k = \pm 1, \pm 2, \ldots \, .\end{aligned}$$

As the mean and acv.f. do not depend on time, the process is second-order stationary. In fact it is clear that the process is also strictly stationary. The ac.f. is given by

$$\rho(k) = \begin{cases} 1 & k = 0 \\ 0 & k = \pm 1, \pm 2, \ldots \end{cases}$$

A purely random process is sometimes called *white noise*, particularly by engineers. Processes of this type are useful in many situations particularly as constituents of more complicated processes such as moving average processes (Section 3.4.3.).

There has been some controversy as to whether it is possible to have a continuous time purely random process. Such a process would have an ac.f. given by

$$\rho(k) = \begin{cases} 1 & k = 0 \\ 0 & k \neq 0 \end{cases}$$

which is a discontinuous function. It can be shown that such a process would have infinite variance, and is a physically unrealizable phenomenon. However, it turns out to be a useful mathematical idealization for some theoretical purposes, and also as an approximation to certain processes which occur in practice, such as voltage fluctuations in a conductor due to thermal noise, and the impulses acting on a particle suspended in fluid along a given axis to produce Brownian motion.

Cox and Miller (1968) point out that a process approximating to white noise may be obtained by considering either

(a) a process in continuous time with ac.f. $\rho(\tau) = e^{-\lambda |\tau|}$
and letting $\lambda \to \infty$, so that the ac.f. decays very quickly;

or

(b) a purely random process in discrete time at intervals Δt and letting $\Delta t \to 0$.

More generally we may regard continuous white noise as any signal made up of the superposition of a large number of independent effects of very brief duration, which appears to behave like continuous white noise when sampled even at quite small discrete intervals.

3.4.2 Random walk Suppose that $\{Z_t\}$ is a discrete, purely random process with mean μ and variance σ_Z^2. A process $\{X_t\}$

is said to be a random walk if

$$X_t = X_{t-1} + Z_t. \tag{3.1}$$

The process is customarily started at zero when $t = 0$ so that

$$X_1 = Z_1$$

and

$$X_t = \sum_{i=1}^{t} Z_i.$$

Then we find $E(X_t) = t\mu$ and that $\text{Var}(X_t) = t\sigma_Z^2$.

As the mean and variance change with t, the process is non-stationary.

However it is interesting to note that the first differences of a random walk, given by

$$Z_t = X_t - X_{t-1}$$

form a purely random process, which is therefore stationary.

The best-known examples of time series which behave very much like random walks are share prices. When these are analysed, a model which often gives a good approximation to the data is

Share price on day t = share price on day $(t - 1)$
+ random error.

3.4.3 Moving average processes Suppose that $\{Z_t\}$ is a purely random process with mean zero and variance σ_Z^2. Then a process $\{X_t\}$ is said to be a moving average process of order q (abbreviated to a MA(q) process) if

$$X_t = \beta_0 Z_t + \beta_1 Z_{t-1} + \ldots + \beta_q Z_{t-q} \tag{3.2}$$

where $\{\beta_i\}$ are constants. The Z's are usually scaled so that $\beta_0 = 1$.

We find immediately that

$$E(X_t) = 0$$

41

and $\mathrm{Var}(X_t) = \sigma_Z^2 \sum_{i=0}^{q} \beta_i^2$

since the Z's are independent. We also have

$$
\begin{aligned}
\gamma(k) &= \mathrm{Cov}\,(X_t, X_{t+k}) \\
&= \mathrm{Cov}\,(\beta_0 Z_t + \ldots + \beta_q Z_{t-q}, \beta_0 Z_{t+k} \\
&\quad + \ldots + \beta_q Z_{t+k-q}) \\
&= \begin{cases}
0 & k > q \\
\sigma_Z^2 \displaystyle\sum_{i=0}^{q-k} \beta_i\,\beta_{i+k} & k = 0, 1, \ldots, q \\
\gamma(-k) & k < 0
\end{cases}
\end{aligned}
$$

since $\mathrm{Cov}(Z_s, Z_t) = \begin{cases} \sigma_Z^2 & s = t \\ 0 & s \neq t \end{cases}$

As $\gamma(k)$ does not depend on t, and the mean is constant, the process is second-order stationary for all values of the $\{\beta_i\}$. Furthermore if the Z's are normally distributed, so are the X's, and we have a normal process completely determined by the mean and acv.f. The process is then strictly stationary.

The ac.f. of the MA(q) process is given by

$$
\rho(k) = \begin{cases}
1 & k = 0 \\
\displaystyle\sum_{i=0}^{q-k} \beta_i\,\beta_{i+k} \Big/ \sum_{i=0}^{q} \beta_i^2 & k = 1, \ldots, q \\
0 & k > q \\
\rho(-k) & k < 0
\end{cases}
$$

Note that the ac.f. 'cuts off' at lag q which is a special feature of MA processes. In particular the MA(1) process with $\beta_0 = 1$ has ac.f.

$$
\rho(k) = \begin{cases}
1 & k = 0 \\
\beta_1/(1 + \beta_1^2) & k = \pm 1 \\
0 & \text{otherwise}
\end{cases}
$$

42

Although no restrictions on the $\{\beta_i\}$ are required for a MA process to be stationary, Box and Jenkins (1970, p. 50) propose restrictions on the $\{\beta_i\}$ to ensure what they call 'invertibility'. This condition is discussed by Chatfield and Prothero (1973b), and arises in the following way. Consider the following first order MA processes:

A) $\qquad X_t = Z_t + \theta Z_{t-1}$

B) $\qquad X_t = Z_t + \dfrac{1}{\theta} Z_{t-1}.$

It can easily be shown that these different processes have exactly the same ac.f. Thus we cannot identify a MA process uniquely from a given ac.f. Now, if we express models A and B by putting Z_t in terms of $X_t, X_{t-1}, \ldots$, we find by successive substitution that

A) $\qquad Z_t = X_t - \theta X_{t-1} + \theta^2 X_{t-2} - \ldots$

B) $\qquad Z_t = X_t - \dfrac{1}{\theta} X_{t-1} + \dfrac{1}{\theta^2} X_{t-2} - \ldots.$

If $|\theta| < 1$, the series for A converges whereas that for B does not. Thus an estimation procedure which involves estimating the residuals (see Section 4.3.1) will lead naturally to model A. Thus if $|\theta| < 1$, model A is said to be invertible whereas model B is not. The imposition of the invertibility condition ensures that there is a unique MA process for a given ac.f.

The invertibility condition for the general order MA process is best expressed by using the backward shift operator, denoted by B, which is defined by

$$B^j X_t = X_{t-j} \qquad \text{for all } j.$$

Then equation (3.2) may be written as

$$X_t = (\beta_0 + \beta_1 B + \ldots + \beta_q B^q) Z_t$$
$$= \theta(B) Z_t$$

where $\theta(B)$ is a polynomial of order q in B. A MA process of order q is invertible if the roots of the equation (regarding B as a complex variable and not an operator)

$$\theta(B) = \beta_0 + \beta_1 B + \ldots + \beta_q B^q = 0$$

all lie outside the unit circle (Box and Jenkins, 1970; p. 50).

MA processes have been used in many areas, particularly econometrics. For example economic indicators are affected by a variety of 'random' events such as strikes, government decisions, shortage of key materials and so on. Such events will not only have an immediate effect but may also affect economic indicators to a lesser extent in several subsequent periods, and so it is at least plausible that a MA process may be appropriate.

Recently Box and Jenkins (1970) have advocated the use of MA processes in conjunction with processes called autoregressive processes which are considered in the next section.

Note that an arbitrary constant, μ say, may be added to the right hand side of equation (3.2) to give a process with mean μ. This does not affect the ac.f. and has been omitted for simplicity.

3.4.4 Autoregressive processes   Suppose that $\{Z_t\}$ is a purely random process with mean zero and variance σ_Z^2. Then a process $\{X_t\}$ is said to be an autoregressive process of order p if

$$X_t = \alpha_1 X_{t-1} + \ldots + \alpha_p X_{t-p} + Z_t. \tag{3.3}$$

This is rather like a multiple regression model, but X_t is regressed not on independent variables but on past values of X_t. Hence the term *auto*regressive. An autoregressive process of order p will be abbreviated to an AR(p) process.

(a) First-order process   For simplicity, we begin by examining the first-order case, where $p = 1$. Then

$$X_t = \alpha X_{t-1} + Z_t \tag{3.4}$$

The AR(1) process is sometimes called the Markov process, after the Russian A. A. Markov. By successive substitution in (3.4) we may write

$$\begin{aligned} X_t &= \alpha[\alpha X_{t-2} + Z_{t-1}] + Z_t \\ &= \alpha^2 [\alpha X_{t-3} + Z_{t-2}] + \alpha Z_{t-1} + Z_t \end{aligned}$$

and eventually we find that X_t may be expressed as an infinite-order MA process in the form (provided $-1 < \alpha < +1$)

$$X_t = Z_t + \alpha Z_{t-1} + \alpha^2 Z_{t-2} + \ldots$$

This duality between AR and MA processes is useful for a variety of purposes. Rather than use successive substitution, it is simpler to use the backward shift operator B. Then equation (3.4) may be written

$$(1 - \alpha B)X_t = Z_t$$

so that

$$\begin{aligned} X_t &= Z_t/(1 - \alpha B) \\ &= (1 + \alpha B + \alpha^2 B^2 + \ldots)Z_t \\ &= Z_t + \alpha Z_{t-1} + \alpha^2 Z_{t-2} + \ldots . \end{aligned}$$

Then we find

$$E(X_t) = 0$$

and

$$\mathrm{Var}(X_t) = \sigma_Z^2 (1 + \alpha^2 + \alpha^4 + \ldots)$$

Thus the variance is finite provided that $|\alpha| < 1$, in which case

$$\mathrm{Var}(X_t) = \sigma_Z^2/(1 - \alpha^2) = \sigma_X^2 .$$

The acv.f. is given by

$$\gamma(k) = E[X_t X_{t+k}]$$
$$= E\{[\Sigma \alpha^i Z_{t-i}][\Sigma \alpha^j Z_{t+k-j}]\}$$
$$= \sigma_Z^2 \sum_{i=0}^{\infty} \alpha^i \alpha^{k+i} \qquad \text{(for } k \geq 0)$$

which converges for $|\alpha| < 1$ to

$$\gamma(k) = \alpha^k \sigma_Z^2 / (1 - \alpha^2)$$
$$= \alpha^k \sigma_X^2$$

For $k < 0$, we find $\gamma(k) = \gamma(-k)$. Since $\gamma(k)$ does not depend on t, an AR process of order 1 is second-order stationary provided that $|\alpha| < 1$. The ac.f. is given by

$$\rho(k) = \alpha^k \qquad (k = 0, 1, 2, \ldots)$$

To get an even function defined for all integer k we take

$$\rho(k) = \alpha^{|k|} \qquad (k = 0, \pm 1, \pm 2, \ldots)$$

The ac.f. may in fact be obtained more simply in the following way, by assuming *a priori* that the process is stationary in which case $E(X_t) = 0$. Multiply through equation (3.4) by X_{t-k} (NOT X_{t+k}!) and take expectations. Then we find, for $k > 0$, that

$$\gamma(-k) = \alpha \gamma(-k + 1)$$

assuming that $E(Z_t X_{t-k}) = 0$ for $k > 0$. Since $\gamma(k)$ is an even function, we must also have

$$\gamma(k) = \alpha \gamma(k - 1) \qquad \text{for } k > 0.$$

Now $\gamma(0) = \sigma_X^2$, and so $\gamma(k) = \alpha^k \sigma_X^2$ for $k \geq 0$. Thus $\rho(k) = \alpha^k$ for $k \geq 0$. Now since $|\rho(k)| \leq 1$, we must have $|\alpha| \leq 1$. But if $|\alpha| = 1$, then $|\rho(k)| = 1$ for all k, which is a degenerate case. Thus $|\alpha| < 1$ for a proper stationary process.

The above method of obtaining the ac.f. is often used,

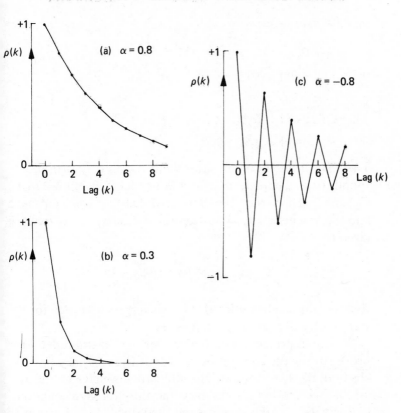

Figure 3.1   Three examples of the autocorrelation function of a first-order autoregressive process with (a) α = 0.8; (b) α = 0.3; (c) α = −0.8.

even though it involves an initial assumption of stationarity.

Three examples of the ac.f. of a first-order AR process are shown in Fig. 3.1 for α = 0.8, 0.3, −0.8. Note how quickly the ac.f. decays when α = 0.3, and note how the ac.f. alternates when α is negative.

(b) General-order case   As in the first-order case, we can express an AR process of finite order as a MA process of infinite order. This may be done by successive substitution, or by using the backward shift operator. Then equation 3.3

47

may be written as

$$(1 - \alpha_1 B - \ldots - \alpha_p B^p)X_t = Z_t$$

or
$$X_t = Z_t/(1 - \alpha_1 B - \ldots - \alpha_p B^p)$$
$$= f(B) Z_t$$

where
$$f(B) = (1 - \alpha_1 B - \ldots - \alpha_p B^p)^{-1}$$
$$= (1 + \beta_1 B + \beta_2 B^2 + \ldots)$$

The relationship between the α's and the β's may then be found. Having expressed X_t as a MA process, it follows that $E(X_t) = 0$. The variance is finite provided that $\Sigma \beta_i^2$ converges, and this is a necessary condition for stationarity. The acv.f. is given by

$$\gamma(k) = \sigma_Z^2 \sum_{i=0}^{\infty} \beta_i \beta_{i+k} \qquad (\text{where } \beta_0 = 1)$$

A sufficient condition for this to converge, and hence for stationarity, is that $\Sigma |\beta_i|$ converges.

We can in principle find the ac.f. of the general-order AR process using the above procedure, but the $\{\beta_i\}$ may be algebraically hard to find. The alternative simpler way is to *assume* the process is stationary, multiply through equation (3.3) by X_{t-k}, take expectations, and divide by σ_X^2 assuming that the variance of X_t is finite. Then, using the fact that $\rho(k) = \rho(-k)$ for all k, we find

$$\rho(k) = \alpha_1 \rho(k-1) + \ldots + \alpha_p \rho(k-p) \qquad \text{for all } k > 0.$$

This set of equations is called the Yule—Walker equations after G. U. Yule and Sir Gilbert Walker. They are a set of difference equations and have the general solution

$$\rho(k) = A_1 \pi_1^{|k|} + \ldots + A_p \pi_p^{|k|}$$

where $\{\pi_i\}$ are the roots of

$$y^p - \alpha_1 y^{p-1} \ldots - \alpha_p = 0.$$

The constants $\{A_i\}$ are chosen to satisfy the initial conditions depending on $\rho(0) = 1$. Thus we have $\Sigma A_i = 1$, and the first $(p - 1)$ Yule–Walker equations provide $(p - 1)$ further restrictions on the $\{A_i\}$ using $\rho(0) = 1$ and $\rho(k) = \rho(-k)$.

From the general form of $\rho(k)$, it is clear that $\rho(k)$ tends to zero as k increases provided that $|\pi_i| < 1$ for all i, and this is a necessary and sufficient condition for the process to be stationary.

An equivalent way of expressing the stationarity condition is to say that the roots of the equation

$$\phi(B) = 1 - \alpha_1 B - \ldots - \alpha_p B^p = 0$$

must lie outside the unit circle (Box and Jenkins, 1970; Section 3.2).

Of particular interest is the AR(2) process, when π_1, π_2 are the roots of the quadratic equation

$$y^2 - \alpha_1 y - \alpha_2 = 0.$$

Thus $|\pi_i| < 1$ if

$$\left| \frac{\alpha_1 \pm \sqrt{(\alpha_1^2 + 4\alpha_2)}}{2} \right| < 1$$

from which it can be shown, (exercise 6), that the stationarity region is the triangular region satisfying

$$\left.\begin{array}{l} \alpha_1 + \alpha_2 < 1 \\ \alpha_1 - \alpha_2 > -1 \\ \alpha_2 > -1 \end{array}\right\}$$

The roots are real if $\alpha_1^2 + 4\alpha_2 > 0$, in which case the ac.f. decreases exponentially with k, but the roots are complex if $\alpha_1^2 + 4\alpha_2 < 0$ in which case we find that the ac.f. is a damped cosine wave. (See for example Cox and Miller, 1968; Section 7.2).

When the roots are real, the constants A_1, A_2 are found as follows. Since $\rho(0) = 1$, we have

$$A_1 + A_2 = 1.$$

From the first of the Yule–Walker equations, we have

$$\rho(1) = \alpha_1 \rho(0) + \alpha_2 \rho(-1)$$
$$= \alpha_1 + \alpha_2 \rho(1)$$

Thus $\rho(1) = \alpha_1 / (1 - \alpha_2)$
$$= A_1 \pi_1 + A_2 \pi_2$$
$$= A_1 \pi_1 + (1 - A_1) \pi_2.$$

Hence we find

$$A_1 = [\alpha_1 / (1 - \alpha_2) - \pi_2] / (\pi_1 - \pi_2)$$

and $A_2 = 1 - A_1$.

AR processes have been applied to many situations in which it is reasonable to assume that the present value of a time series depends on the immediate past values together with a random error. For simplicity we have only considered processes with mean zero, but non-zero means may be dealt with by rewriting equation 3.3 in the form

$$X_t - \mu = \alpha_1 (X_{t-1} - \mu) + \ldots + \alpha_p (X_{t-p} - \mu) + Z_t.$$

This does not affect the ac.f.

3.4.5 Mixed models A useful class of models for time series is that formed from a combination of MA and AR processes. A mixed autoregressive-moving average process containing p AR terms and q MA terms (abbreviated to an ARMA (p, q) process) is said to be of order (p, q). It is given by

$$X_t = \alpha_1 X_{t-1} + \ldots + \alpha_p X_{t-p} + Z_t + \beta_1 Z_{t-1} \\ + \ldots + \beta_q Z_{t-q} \tag{3.5}$$

Using the backward shift operator B, equation (3.5) may be written in the form

$$\phi(B)X_t = \theta(B)Z_t \qquad (3.5a)$$

where $\phi(B)$, $\theta(B)$ are polynomials of order p, q respectively, such that

$$\phi(B) = 1 - \alpha_1 B - \ldots - \alpha_p B^p$$

and

$$\theta(B) = 1 + \beta_1 B + \ldots + \beta_q B^q$$

The values of α_i which make the process stationary are such that the roots of

$$\phi(B) = 0$$

lie outside the unit circle. The values of $\{\beta_i\}$ which make the process invertible are such that the roots of

$$\theta(B) = 0$$

lie outside the unit circle.

It is fairly straightforward, though algebraically rather tedious, to calculate the ac.f. of an ARMA process, but this will not be discussed here. (See exercise 11 and Box and Jenkins, 1970; Section 3.4).

The importance of ARMA processes lies in the fact that a stationary time series may often be described by an ARMA model involving fewer parameters than a MA or AR process by itself.

3.4.6 Integrated models   In practice most time series are non-stationary. In order to fit a stationary model, such as those discussed in Sections 3.4.3–3.4.5, it is necessary to remove non-stationary sources of variation. If the observed time series is non-stationary in the mean then we can difference the series, as suggested in Section 2.5.3, and this approach is widely used in econometrics. If X_t is replaced by $\nabla^d X_t$ in equation 3.5 then we have a model capable of describing certain types of non-stationary series. Such a model is called an 'integrated' model because the stationary model which is fitted to the differenced data has to be summed or 'integrated' to provide a model for the non-stationary data. Writing

$$W_t = \nabla^d X_t,$$

the general autoregressive integrated moving average process (abbreviated ARIMA process) is of the form

$$W_t = \alpha_1 W_{t-1} + \ldots + \alpha_p W_{t-p} + Z_t + \ldots + \beta_q Z_{t-q}$$
(3.6)

The above ARIMA process describing the dth differences of the data is said to be of order (p, d, q). In practice the value of d is often taken to be one. Note that the random walk can be regarded as an ARIMA $(0, 1, 0)$ process.

3.4.7 The general linear process The infinite-order MA process with non-zero mean given by

$$X_t - \mu = \sum_{i=0}^{\infty} \beta_i Z_{t-i}$$
(3.7)

is often called the general linear process since processes of this type can be obtained by passing a purely random process through a linear system (see Chapter 9). Both MA and AR processes are special cases of the general linear process and the duality between these two classes is easily demonstrated using the backward shift operator. A MA process of finite order can be expressed as an AR process of infinite order, while an AR process of finite order can be expressed as a MA process of infinite order (see exercises 9 and 10).

†3.4.8 Continuous processes So far, we have only considered stochastic processes in discrete time, because these are the

† This section may be omitted at first reading.

main type of process the statistician uses in practice. Continuous processes have been used in some applications, notably in the study of control theory by electrical engineers. Here we shall only indicate some of the problems connected with their use.

By analogy with equation (3.7), we could try to define the general linear process in continuous time by

$$X(t) = \int_0^\infty h(u) \, Z(t - u) \, du, \tag{3.8}$$

but if $Z(t)$ is continuous white noise, there are immediate mathematical difficulties regarding the above integral (Cox and Miller, 1968; Section 7.4). As discussed in Section 3.4.1, continuous white noise is not physically realizable and so equation (3.8) is not generally used. Most authors, such as Yaglom (1962; p. 22), only consider continuous processes which have a continuous ac.f. However Jenkins and Watts (1968) do use equation (3.8).

As an example of a continuous process, we will briefly consider a first-order continuous AR process. A first-order discrete AR process (equation 3.4) may be written in terms of X_t, ∇X_t and Z_t. As differencing in discrete time corresponds to differentiation in continuous time, a natural way of defining a first-order AR process is by

$$\frac{dX(t)}{dt} + aX(t) = Z(t) \tag{3.9}$$

which is essentially the Langevin equation (Cox and Miller, 1968; p. 298). Yaglom (1962; p. 69) suggests that it is more legitimate to write equation (3.9) in the form

$$dX(t) + aX(t)dt = dU(t) \tag{3.10}$$

where $\{U(t)\}$ is a process with orthogonal increments (i.e. the random variables $[U(t_2) - U(t_1)]$ and $[U(t_4) - U(t_3)]$ are uncorrelated for any two non-over-lapping intervals

(t_1, t_2) and (t_3, t_4)). Equation (3.10) is the same as that which defines the Ornstein–Uhlenbeck process (Cox and Miller, 1968; Section 5.8). It can be shown that the process $X(t)$ defined by equation (3.10) has ac.f.

$$\rho(u) = e^{-a|u|}$$

which is similar to the ac.f. of a first-order discrete AR process in that both decay exponentially.

The rigorous study of continuous processes requires considerable mathematical skill, including a knowledge of stochastic integration, and we will not pursue it here.

†3.5 The Wold decomposition theorem

This section gives a brief introduction to a famous theorem given by H. Wold in 1938 which applies to any stationary process in discrete time. The treatment in this section is a shortened version of that given by Cox and Miller (1968). The Wold decomposition theorem says that any discrete stationary process can be expressed as the sum of two uncorrelated processes, one purely deterministic and one which is purely indeterministic. The terms deterministic and indeterministic are defined as follows. We can regress X_t on $(X_{t-q}, X_{t-q-1} \ldots)$ and denote the residual variance from the resulting linear regression model by τ_q^2. As $\tau_q^2 \leqslant \mathrm{Var}(X_t)$, it is clear that, as q increases, τ_q^2 is a non-decreasing bounded sequence and therefore tends to a limit as $q \to \infty$. If $\lim_{q \to \infty} \tau_q^2 = \mathrm{Var}(X_t)$, then linear regression on the remote past is useless for prediction purposes and we say that $\{X_t\}$ is *purely indeterministic*. But if $\lim_{q \to \infty} \tau_q^2$ is zero, then the process can be forecast exactly, and we say that $\{X_t\}$ is purely deterministic.

All the stationary processes we have considered in this chapter, such as AR and MA processes, are purely indeterministic. The best-known examples of purely

† This section may be omitted at first reading.

eterministic processes are sinusoidal processes (see exercise
4) such as

$$X_t = g \cos(\omega t + \theta) \qquad (3.11)$$

here g is a constant, ω is a constant in $(0, \pi)$ called the
equency of the process and θ is a random variable called the
hase which is uniformly distributed on $(0, 2\pi)$ but which is
xed for a single realization. Note that we must include the
erm θ so that

$$E(X_t) = 0 \qquad \text{for all } t,$$

therwise (3.11) would not define a stationary process. As θ
 fixed for a single realization, once enough values of X_t have
een observed to evaluate θ, all subsequent values of X_t are
ompletely determined. It is then obvious that (3.11) defines
 purely deterministic process.

The Wold decomposition theorem also says that the purely
ideterministic component can be written as the linear sum
f an 'innovation' process $\{Z_t\}$, which is a sequence of
ncorrelated random variables. A special class of processes of
articular interest arise when the Z's are independent and not
ierely uncorrelated, as we then have a general linear process
Section 3.4.7). On the other hand when processes are
enerated in a non-linear way the Wold decomposition is
sually of little interest.

The concept of a purely indeterministic process is a useful
ne and most of the stationary stochastic processes which are
onsidered in the rest of this book are of this type.

Exercises

n all the following questions $\{Z_t\}$ is a discrete, purely
andom process, such that $E(Z_t) = 0$, $V(Z_t) = \sigma_Z^2$,
$\text{Cov}(Z_t, Z_{t+k}) = 0$, $k \neq 0$.

1. Find the ac.f. of the second-order MA process given by
$$X_t = Z_t + 0.7Z_{t-1} - 0.2Z_{t-2}.$$

2. Show that the ac.f. of the mth-order MA process given by
$$X_t = \sum_{k=0}^{m} Z_{t-k}/(m+1)$$
is
$$\rho(k) = \begin{cases} (m+1-k)/(m+1) & k = 0, 1, \ldots, m. \\ 0 & k > m \end{cases}$$

3. Show that the infinite-order MA process $\{X_t\}$ defined by
$$X_t = Z_t + C(Z_{t-1} + Z_{t-2} + \ldots)$$
where C is a constant, is non-stationary.
Also show that the series of first differences $\{Y_t\}$ defined by
$$Y_t = X_t - X_{t-1}$$
is a first-order MA process and is stationary. Find the ac.f. of $\{Y_t\}$

4. Find the ac.f. of the first-order AR process defined by
$$X_t = 0.7X_{t-1} + Z_t.$$
Plot $\rho(k)$ for $k = -6, -5, \ldots, -1, 0, +1, \ldots, +6$.

5. If $X_t = \mu + Z_t + \beta Z_{t-1}$, where μ is a constant, show that the ac.f. does not depend on μ.

6. Find the values of λ_1, λ_2, such that the second-order AR process defined by
$$X_t = \lambda_1 X_{t-1} + \lambda_2 X_{t-2} + Z_t$$
is stationary.

If $\lambda_1 = \frac{1}{3}$, $\lambda_2 = \frac{2}{9}$, show that the ac.f. of X_t is given by

$$\rho(k) = \frac{16}{21} \left(\frac{2}{3}\right)^{|k|} + \frac{5}{21} \left(-\frac{1}{3}\right)^{|k|}.$$

$$k = 0, \pm1, \pm2, \ldots$$

7. Explain what is meant by a weakly (or second-order) stationary process, and define the ac.f. $\rho(u)$ for such a process. Show that $\rho(u) = \rho(-u)$ and that $|\rho(u)| \leqslant 1$.

Show that the ac.f. of the stationary second-order AR process,

$$X_t = \frac{1}{12}X_{t-1} + \frac{1}{12}X_{t-2} + Z_t$$

is given by $\rho(k) = \dfrac{45}{77}\left(\dfrac{1}{3}\right)^{|k|} + \dfrac{32}{77}\left(-\dfrac{1}{4}\right)^{|k|}.$

$$k = 0, \pm1, \pm2, \ldots$$

8. The stationary process $\{X_t\}$ has acv.f. $\gamma_X(k)$. A new stationary process $\{Y_t\}$ is defined by $Y_t = X_t - X_{t-1}$. Obtain the acv.f. of $\{Y_t\}$ in terms of $\gamma_X(k)$ and find $\gamma_Y(k)$ when $\gamma_X(k) = \lambda^{|k|}$.

9. Find the MA representation of the first-order AR process

$$X_t = 0.3X_{t-1} + Z_t.$$

10. Suppose that a stationary process $\{X_t\}$ can be represented in two equivalent forms by

$$a_0 X_t + a_1 X_{t-1} + \ldots = Z_t$$

and

$$X_t = b_0 Z_t + b_1 Z_{t-1} + \ldots$$

where we may take $a_0 = b_0 = 1$. These are the AR and

MA representations of the process. Let

$$A(s) = \sum_{i=0}^{\infty} a_i s^i$$

and $B(s) = \sum_{i=0}^{\infty} b_i s^i$. Show that $A(s) B(s) = 1$.

By defining $\Gamma(s) = \sum_{k=-\infty}^{\infty} \gamma_X(k) s^k$ also show that

$$\Gamma(s) = \sigma_Z^2 \, B(s) \, B\left(\frac{1}{s}\right) = \sigma_Z^2 \left/ \left[A(s) \, A\left(\frac{1}{s}\right) \right] \right. .$$

11. Show that the ac.f. of the mixed ARMA model

$$X_t = \alpha X_{t-1} + Z_t + \beta Z_{t-1}$$

is given by

$$\rho(1) = (1 + \alpha\beta)(\alpha + \beta)/(1 + \beta^2 + 2\alpha\beta)$$
$$\rho(k) = \alpha\rho(k-1) \qquad (k = 2, 3, \ldots)$$

12. A Markov process $X(t)$ in continuous time has two possible states, 0 and 1. Let

$$P[X(t + \Delta t) = 1 \mid X(t) = 0] = \alpha\Delta t + o(\Delta t)$$
$$P[X(t + \Delta t) = 0 \mid X(t) = 1] = \beta\Delta t + o(\Delta t)$$

It can be shown that

$$P[X(t) = 0] = \beta/(\alpha + \beta) + [P(X(0) = 0)$$
$$- \beta/(\alpha + \beta)] \, e^{-(\alpha + \beta)t}$$

If $P(X(0) = 0) = \beta/(\alpha + \beta)$ show that the process is second-order stationary.
(Hint: Show that $E(X(t) X(t + \tau)) =$

$$\frac{\alpha}{\alpha + \beta} \left[\alpha/(\alpha + \beta) + (1 - \alpha/(\alpha + \beta)) \, e^{-(\alpha + \beta)\tau} \right]$$

is a function of τ only.)

13. Show that the AR(2) process

$$X_t = X_{t-1} + cX_{t-2} + Z_t$$

is stationary provided $-1 < c < 0$. Find the autocorrelation function when $c = -3/16$.

Show that the AR(3) process

$$X_t = X_{t-1} + cX_{t-2} - cX_{t-3} + Z_t$$

is non-stationary for all values of c.

14. For a complex-valued process $X(t)$, with (complex) mean μ, the acv.f. is defined by

$$\gamma(\tau) = E[(X(t) - \mu)(\overline{X}(t+\tau) - \overline{\mu})]$$

where the overbar denotes the complex conjugate. Show that the process $X(t) = Ye^{i\omega t}$ is second-order stationary where Y is a complex random variable mean zero which does not depend on t, and ω is a real constant. One useful form for the random variable Y occurs when it takes the form $ge^{i\theta}$ where g is a constant and θ is a uniformly distributed random variable on $(0, 2\pi)$. Show that $E(Y) = 0$ in this case (see Yaglom, 1962; Section 2.8, but note that the autocovariance function is called the correlation function by Yaglom).

CHAPTER 4

Estimation in the Time Domain

In Chapter 3 we introduced several different types of probability model which may be used to describe time series. In this chapter we discuss the problem of fitting a suitable model to an observed time series, confining ourselves to the discrete-time case. The major diagnostic tool in this chapter is the sample autocorrelation function. Inference based on this function is often called an analysis in the *time domain*.

4.1 Estimating the autocovariance and autocorrelation functions

We have already noted in Section 3.3 that the theoretical ac.f. is an important tool for describing the properties of a stationary stochastic process. In Section 2.7, we heuristically introduced the sample ac.f. of an observed time series, and this is an intuitively reasonable estimate of the theoretical ac.f., provided the series is stationary. This section investigates the properties of the sample ac.f. more closely.

Let us look first at the autocovariance function (acv.f.). The sample autocovariance coefficient at lag k (see equation (2.6)) given by

$$c_k = \sum_{t=1}^{N-k} (x_t - \bar{x})(x_{t+k} - \bar{x})/N \qquad (4.1)$$

is the usual estimator for the theoretical autocovariance coefficient at lag k, $\gamma(k)$. The properties of this estimator are

discussed by Jenkins and Watts (1968, Section 5.3.3) and
Priestley (1981, Chapter 5). It can be shown that the bias in
c_k is of order $1/N$. However

$$\lim_{N \to \infty} E(c_k) = \gamma(k),$$

so that the estimator is asymptotically unbiased.

It can also be shown that

$$\text{Cov}(c_k, c_m) \simeq \sum_{r = -\infty}^{\infty} \{\gamma(r)\gamma(r + m - k) \\ + \gamma(r + m)\gamma(r - k)\}/N \quad (4.2)$$

When $m = k$, formula (4.2) gives us the variance of c_k and
hence the mean square error of c_k. Formula (4.2) also
highlights the fact that successive values of c_k may be highly
correlated and this increases the difficulty of interpreting the
correlogram.

Jenkins and Watts (1968, Chapter 5) compare the
estimator (4.1) with the alternative estimator

$$c'_k = \sum_{t = 1}^{N - k} (x_t - \bar{x})(x_{t+k} - \bar{x})/(N - k).$$

This is used by some authors because it has a smaller bias, but
Jenkins and Watts conjecture that it generally has a higher
mean square error.

A third method of estimating the acv.f. is to use
Quenouille's method of bias reduction, otherwise known as
jackknife estimation. In this procedure the time series is
divided into two halves, and the sample acv.f. is estimated
from each half of the series and also from the whole series.
If the three resulting estimates of $\gamma(k)$ are denoted by c_{k1},
c_{k2} and c_k in an obvious notation, then the jackknife
estimate is given by

$$\tilde{c}_k = 2c_k - \tfrac{1}{2}(c_{k1} + c_{k2}) \quad (4.3)$$

It can be shown that this estimator reduces the bias from order $1/N$ to order $1/N^2$. It has an extra advantage in that one can see if both halves of the time series have similar properties and hence see if the time series is stationary. However, the method has the disadvantage that it requires extra computation. It is also sensitive to non-stationarity in the mean and c_{k1}, c_{k2}, should be compared with the overall c_k as well as with each other.

Having estimated the acv.f., we then take

$$r_k = c_k/c_0 \tag{4.4}$$

as an estimator for $\rho(k)$. The properties of r_k are rather more difficult to find than those of c_k because it is the ratio of two random variables. It can be shown that r_k is generally biased. The bias can be reduced by jackknifing as in equation (4.3). The jackknife estimator is given in an obvious notation by

$$\tilde{r}_k = 2r_k - \tfrac{1}{2}(r_{k1} + r_{k2})$$

A general formula for the variance of r_k is given by Kendall and Stuart (1966) and depends on *all* the autocorrelation coefficients of the process. We will only consider the properties of r_k when sampling from a purely random process, when all the theoretical autocorrelation coefficients are zero except at lag zero. These results are the most useful since they enable us to decide if the observed values of r_k from a given time series are significantly different from zero.

Suppose that $x_1, \ldots, x_N$ are independent and identically distributed random variables with arbitrary mean. Then it can be shown (Kendall and Stuart, 1966, Chapter 48) that

$$E(r_k) \simeq -1/N$$

and

$$\mathrm{Var}(r_k) \simeq 1/N$$

and that r_k is asymptotically normally distributed under weak conditions. Thus having plotted the correlogram, as described in Section 2.7, we can plot approximate 95% confidence limits at $-1/N \pm 2/\sqrt{N}$, which are often further approximated to $\pm 2/\sqrt{N}$. Observed values of r_k which fall outside these limits are 'significantly' different from zero at the 5% level. However, when interpreting a correlogram, it must be remembered that the overall probability of getting a coefficient outside these limits, given that the data really are random, increases with the number of coefficients plotted. For example if the first 20 values of r_k are plotted then one expects one 'significant' value even if the data are random. Thus, if only one or two coefficients are 'significant', the size and lag of these coefficients must be taken into account when deciding if a set of data is random. Values well outside the 'null' confidence limits indicate non-randomness. So also do significant coefficients at a lag which has some physical interpretation, such as lag one or a lag corresponding to seasonal variation.

Fig. 4.1 shows the correlogram for 100 observations, generated on a computer, which are supposed to be independent normally distributed variables. The confidence limits are approximately $\pm 2/\sqrt{100} = \pm 0.2$. We see that 2 of the first 20 values of r_k lie just outside the significance limits. As these occur at apparently arbitrary lags (namely 12 and 17) we conclude that there is no firm evidence to disprove the hypothesis that the observations are independently distributed.

4.1.1 Interpreting the correlogram   We have already given some general advice on interpreting correlograms in Section 2.7.2. The correlogram is also helpful in identifying which type of ARIMA model gives the best representation of an observed time series. A correlogram like that in Fig. 2.3, where the values of r_k do not come down to zero reasonably quickly, indicates non-stationarity and so the series needs to

Figure 4.1   The correlogram of 100 'independent' normally distributed observations.

be differenced. For stationary series, the correlogram is compared with the theoretical ac.f.s of different ARMA processes in order to choose the one which is most appropriate. The ac.f. of a MA(q) process is easy to recognize as it 'cuts off' at lag q whereas the ac.f. of an AR(p) process is a mixture of damped exponentials and sinusoids and dies out slowly (or attenuates). The ac.f. of a mixed ARMA model will also generally attenuate rather than 'cut off'. The partial autocorrelation function (see Section 4.2.2) or the inverse autocorrelation function (see Chapter 10) should also be examined as they 'cut off' for AR models and attenuate for MA models and give further aid in identification.

For example, if we find that r_1 is significantly different from zero but that subsequent values of r_k are all close to zero then an MA(1) model is indicated since its theoretical ac.f. is of this form. Alternatively, if $r_1, r_2, r_3, \ldots$ appear to be decreasing exponentially, then an AR(1) model may be appropriate.

The interpretation of correlograms is one of the hardest aspects of time-series analysis and practical experience is a 'must'.

4.1.2 Ergodic theorems   The fact that one can obtain consistent estimates of the properties of a stationary process from a single finite realization is not immediately obvious. However some theorems, called ergodic theorems, have been proved which show that, for most stationary processes which are likely to be met in practice, the sample moments of an observed record of length T converge (in mean square) to the corresponding population moments as $T \rightarrow \infty$. In other words time averages for a single realization converge to ensemble averages. See for example Yaglom (1962, Section 1.4). We will not pursue the topic here.

4.2 Fitting an autoregressive process

Having estimated the ac.f. of a given time series, we should have some idea as to what stochastic process will provide a suitable model. If an AR process is thought to be appropriate, there are two related questions:

(a) What is the order of the process?
(b) How can we estimate the parameters of the process?

We will consider question (b) first.

4.2.1. Estimating the parameters of an autoregressive process
Suppose we have an AR process of order p, with mean μ, given by

$$X_t - \mu = \alpha_1 (X_{t-1} - \mu) + \ldots + \alpha_p (X_{t-p} - \mu) + Z_t$$

$$(4.5)$$

Given N observations, $x_1, \ldots, x_N$, the parameters μ, $\alpha_1, \ldots, \alpha_p$, may be estimated by least squares by minimising

$$S = \sum_{t=p+1}^{N} [x_t - \mu - \alpha_1(x_{t-1} - \mu) - \ldots \\ -\alpha_p(x_{t-p} - \mu)]^2$$

with respect to $\mu, \alpha_1, \ldots, \alpha_p$. If the Z_t process is normal, then the least squares estimates are in addition maximum likelihood estimates (Jenkins and Watts, 1968, Section 5.4) conditional on the first p values in the time series being fixed.

In the first-order case, with $p = 1$, we find (see Exercise 1),

$$\hat{\mu} = \frac{\bar{x}_{(2)} - \hat{\alpha}_1 \bar{x}_{(1)}}{1 - \hat{\alpha}_1} \tag{4.6}$$

and

$$\hat{\alpha}_1 = \frac{\sum_{t=1}^{N-1} (x_t - \hat{\mu})(x_{t+1} - \hat{\mu})}{\sum_{t=1}^{N-1} (x_t - \hat{\mu})^2} \tag{4.7}$$

where $\bar{x}_{(1)}, \bar{x}_{(2)}$ are the means of the first and last $(N - 1)$ observations. Now since

$$\bar{x}_{(1)} \simeq \bar{x}_{(2)} \simeq \bar{x}$$

we have approximately that

$$\hat{\mu} \simeq \bar{x}. \tag{4.8}$$

This approximate estimator, which is intuitively appealing, is nearly always used in preference to (4.6). Substituting this value into (4.7) we have

$$\hat{\alpha}_1 = \frac{\sum_{t=1}^{N-1} (x_t - \bar{x})(x_{t+1} - \bar{x})}{\sum_{t=1}^{N-1} (x_t - \bar{x})^2}. \tag{4.9}$$

It is interesting to note that this is exactly the same estimator that would arise if we were to treat the autoregressive equation

$$X_t - \bar{x} = \alpha_1 (x_{t-1} - \bar{x}) + Z_t$$

as an ordinary regression with $(x_{t-1} - \bar{x})$ as the 'independent' variable. In fact H. B. Mann and A. Wald showed in 1943 that, asymptotically, much of classical regression theory can be applied to autoregressive situations.

A further approximation which is often used is obtained by noting that the denominator of (4.9) is approximately

$$\sum_{t=1}^{N} (x_t - \bar{x})^2,$$

so that

$$\hat{\alpha}_1 \simeq c_1 / c_0$$
$$= r_1.$$

This approximate estimator for $\hat{\alpha}_1$ is also intuitively appealing since r_1 is an estimator for $\rho(1)$ and $\rho(1) = \alpha_1$ for a first-order AR process. A confidence interval for α_1 may be obtained from the fact that the asymptotic standard error of $\hat{\alpha}_1$ is $\sqrt{\{(1 - \alpha_1^2)/N\}}$ although the confidence interval will not be symmetric for $\hat{\alpha}_1$ away from zero (Bartlett, 1966). In particular a test for $\alpha_1 = 0$ may be obtained, as suggested by the results of Section 4.1, by seeing if $\hat{\alpha}_1$ lies within the range $\pm 2/\sqrt{N}$.

For a second-order AR process, with $p = 2$, similar approximations may be made to give

$$\hat{\mu} \simeq \bar{x}$$
$$\hat{\alpha}_1 \simeq r_1 (1 - r_2)/(1 - r_1^2) \qquad (4.10)$$
$$\hat{\alpha}_2 \simeq (r_2 - r_1^2)/(1 - r_1^2) \qquad (4.11)$$

These results are also intuitively reasonable in that if we fit a second-order model to what is really a first order process, then as $\alpha_2 = 0$, we have $\rho(2) = \rho(1)^2 = \alpha_1^2$ and so $r_2 \simeq r_1^2$. Thus equations (4.10) and (4.11) become $\hat{\alpha}_1 \simeq r_1$ and $\hat{\alpha}_2 \simeq 0$. Jenkins and Watts (1968; p. 197) describe $\hat{\alpha}_2$ as the (sample) *partial* autocorrelation coefficient of order 2 which measures the excess correlation between $\{X_t\}$ and $\{X_{t+2}\}$ not accounted for by $\hat{\alpha}_1$.

In addition to point estimates of α_1 and α_2 it is also possible to find a confidence region in the (α_1, α_2) plane (Jenkins and Watts, 1968, p. 192).

Higher order AR processes may also be fitted by least squares in a straightforward way. Two alternative approximate methods are commonly used. Both methods involve taking

$$\hat{\mu} = \bar{x}.$$

The first method fits the data to the model

$$X_t - \bar{x} = \alpha_1(x_{t-1} - \bar{x}) + \ldots + \alpha_p(x_{t-p} - \bar{x}) + Z_t$$

treating it as if it were an ordinary regression model. A standard multiple regression computer program may be used with appropriate modification.

The second method involves substituting the sample autocorrelation coefficients into the first p Yule–Walker equations (see Section 3.4.4) and solving for $(\hat{\alpha}_1, \ldots, \hat{\alpha}_p)$ (e.g. Pagano, 1972). In matrix form these equations are

$$\mathbf{R}\hat{\mathbf{a}} = \mathbf{r} \tag{4.12}$$

where

$$\mathbf{R} = \begin{pmatrix} 1 & r_1 & r_2 & \ldots & r_{p-1} \\ r_1 & 1 & r_1 & \ldots & r_{p-2} \\ r_2 & r_1 & 1 & \ldots & r_{p-3} \\ \hdotsfor{5} \\ r_{p-1} & r_{p-2} & & \ldots & 1 \end{pmatrix}$$

is a $(p \times p)$ matrix

$$\hat{\boldsymbol{\alpha}}^T = (\hat{\alpha}_1, \ldots, \hat{\alpha}_p)$$

and

$$\mathbf{r}^T = (r_1, \ldots, r_p).$$

For N reasonably large, both methods will give estimated values 'very close' to the true least squares estimates for which $\hat{\mu}$ is close to but not necessarily equal to $\bar{x}$.

4.2.2 Determining the order of an autoregressive process

It is usually difficult to assess the order of an AR process from the sample ac.f. alone. For a first-order process the theoretical ac.f. decreases exponentially and the sample function should have a similar shape. But for higher order processes the ac.f. may be a mixture of damped exponential or sinusoidal functions and is difficult to identify. One approach is to fit AR processes of progressively higher order, to calculate the residual sum of squares for each value of p, and to plot this against p. It may then be possible to see the value of p where the curve 'flattens out' and the addition of extra parameters gives little improvement in fit.

Another aid to determining the order of an AR process is the *partial* autocorrelation function (see Box and Jenkins, 1970, p. 64) which is defined as follows. When fitting an AR(p) model, the last coefficient, α_p will be denoted by π_p and measures the excess correlation at lag p which is not accounted for by an AR($p - 1$) model. It is called the pth partial autocorrelation coefficient and when plotted against p gives the partial ac.f. Values of π_p which are outside the range $\pm 2/\sqrt{N}$ are significantly different from zero at the 5% level. It can be shown that the partial ac.f. of an AR(p) process 'cuts off' at lag p so that the 'correct' order is assessed as that value of p beyond which the sample values of $\{\pi_j\}$ are not significantly different from zero. Two recent alternative methods for identifying the correct order, based on the *inverse* ac.f. and on Akaike's information criterion, are introduced in Chapter 10.

4.3 Fitting a moving average process

Suppose now that a MA process is thought to be an appropriate model for a given time series. As for an AR process, we have two problems:

(a) Finding the order of the process,

and

(b) Estimating the parameters of the process.

We consider problem (b) first.

4.3.1 Estimating the parameters of a moving average process

Estimation problems are more difficult for an MA process than an AR process, because efficient explicit estimators cannot be found. Instead some form of numerical iteration must be performed.

Let us begin by considering the first-order MA process

$$X_t = \mu + Z_t + \beta_1 Z_{t-1} \tag{4.13}$$

where μ, β_1 are constants and Z_t denotes a purely random process. We would like to write the residual sum of squares, ΣZ_t^2, solely in terms of the observed x's and the parameters μ, β_1, as we did for the AR process, differentiate with respect to μ and β_1 and hence find the least-squares estimates. Unfortunately the residual sum of squares is not a quadratic function of the parameters and so explicit least-squares estimates cannot be found. Nor can we simply equate sample and theoretical first-order autocorrelation coefficients by

$$r_1 = \hat{\beta}_1 / (1 + \hat{\beta}_1^2) \tag{4.14}$$

and choose the solution $\hat{\beta}_1$ such that $|\hat{\beta}_1| < 1$, because this gives rise to an inefficient estimator. (Whittle, 1953).

The approach suggested by Box and Jenkins (1970,

Chapter 7) is as follows. Select suitable starting values for μ and β_1 such as $\mu = \bar{x}$ and β_1 given by the solution of equation 4.14. (See Table A in Box and Jenkins, 1970). Then the corresponding residual sum of squares may be calculated using (4.13) recursively in the form

$$Z_t = X_t - \mu - \beta_1 Z_{t-1} \tag{4.15}$$

With $z_0 = 0$, we have $z_1 = x_1 - \mu, z_2 = x_2 - \mu - \beta_1 z_1, \ldots,$

$$z_N = x_N - \mu - \beta_1 z_{N-1}.$$

Then $\sum\limits_{t=1}^{N} z_t^2$ may be calculated.

This procedure can then be repeated for other values of μ and β_1 and the sum of squares Σz_t^2 computed for a grid of points in the (μ, β_1) plane. We may then determine by inspection the least-squares estimates of μ and β_1 which minimize Σz_t^2. These least-squares estimates are also maximum likelihood estimates conditional on a fixed value of z_0 provided that Z_t is normally distributed. The procedure can be further refined by *back forecasting* the value of z_0 (see Box and Jenkins, 1970) but this is unnecessary except when β_1 is 'close' to plus or minus one.

An alternative estimation procedure due to J. Durbin is to fit a high order AR process to the data and use the duality between AR and MA processes (see for example Kendall, 1976, Section 12.8). This procedure has the advantage of requiring less computation, but the widespread availability of high-speed computers has resulted in the procedure becoming obsolete.

For higher-order processes a similar type of iterative procedure to that described above may be used. For example with a second-order MA process, one would guess starting values for μ, β_1, β_2, compute the residuals recursively using

$$z_t = x_t - \mu - \beta_1 z_{t-1} - \beta_2 z_{t-2}$$

71

and compute Σz_t^2. Then other values of μ, β_1, β_2 would be tried, probably over a grid of points, until the minimum value of Σz_t^2 is found. Clearly a computer is essential for performing such a large number of arithmetic operations, and for processes higher than second order some form of hill-climbing procedure may be desirable to minimize the residual sum of squares in a way that is numerically efficient. Box and Jenkins (1970, Section 7.2) describe such a procedure which they call 'non-linear estimation'. This description arises from the fact that the residuals are non-linear functions of the parameters but the description may give rise to confusion.

For a completely new set of data, it is usually a good idea to use the graphical method based on evaluating the residual sum of squares at a grid of points. A visual examination of the sum of squares surface will sometimes provide useful information. In particular it is interesting to see how 'flat' the surface is; if the surface is approximately quadratic; and if the parameter estimates are approximately uncorrelated.

In addition to point estimates, an approximate confidence region for the model parameters may be found as described by Box and Jenkins (1970, p. 228) by assuming that the Z_t are normally distributed. But there is some doubt as to whether the asymptotic normality of maximum likelihood estimators will apply even for moderately large sample sizes (e.g. $N = 200$).

It should now be clear that it is much harder to estimate the parameters of a MA model than an AR model, as the 'errors' in a MA model are non-linear functions of the parameters and iterative methods are required to minimize the residual sum of squares. Because of this, many analysts prefer to fit an AR model to a given time series even though the resulting model may contain more parameters than the 'best' MA model. Indeed the simplicity of AR modelling is the main reason for their use in the step-wise autoregression forecasting technique (see Section 5.2.5) and in autoregressive spectrum estimation (see Chapter 10).

4.3.2 Determining the order of a moving average process

If a MA process is thought to be appropriate for a given set of data, the order of the process is usually evident from the sample ac.f. The theoretical ac.f. of a MA(q) process has a very simple form in that it 'cuts off' at lag q (see Section 3.4.2) and so the analyst should look for the lag beyond which the values of r_k are close to zero. The more complicated procedure described by Jenkins and Watts (1968, Section 5.4.4) is usually unnecessary.

4.4 Estimating the parameters of a mixed model

Suppose now that a mixed autoregressive-moving average (ARMA) model is thought to be appropriate for a given time series. The estimation problems for an ARMA model are similar to those for a MA model in that an iterative procedure has to be used. The residual sum of squares can be calculated at every point on a suitable grid of the parameter values, and the values which give the minimum sum of squares may then be assessed. Alternatively some sort of optimization procedure may be used.

As an example consider the ARMA(1,1) process whose ac.f. decreases exponentially after lag one (see p. 58). This model may be recognized as appropriate if the sample ac.f. has a similar form. The model is given by

$$X_t - \mu = \alpha_1 (X_{t-1} - \mu) + Z_t + \beta_1 Z_{t-1}.$$

Given N observations $x_1, \ldots, x_N$, we guess values for μ, α_1, β_1, set $z_0 = 0$ and $x_0 = \mu$, and then calculate the residuals recursively by

$$z_1 = x_1 - \mu$$

$$z_2 = x_2 - \mu - \alpha_1 (x_1 - \mu) - \beta_1 z_1$$

. . . .

$$z_N = (x_N - \mu) - \alpha_1 (x_{N-1} - \mu) - \beta_1 z_{N-1}$$

The residual sum of squares $\sum\limits_{t=1}^{N} z_t^2$ may then be calculated.

Then other values of μ, α_1, β_1 may be tried until the minimum residual sum of squares is found. Further details may be found in Box and Jenkins (1970). Many modifications have been suggested for the estimation procedure for ARMA models; see for example, Ljung and Box (1979), Nicholls (1977) and Priestley (1981, Chapter 5).

4.5 Estimating the parameters of an integrated model

In practice, most time series are non-stationary and the stationary models we have so far considered are not immediately appropriate. Then we can difference an observed time series until it is stationary as described in Section 3.4.6. An AR, MA or mixed model may then be fitted to the differenced series as described in Sections 4.2 – 4.4. The resulting model for the undifferenced series is called an integrated (or ARIMA) model.

4.6 The Box–Jenkins seasonal model

In practice, many time series contain a seasonal periodic component which repeats every s observations. For simplicity we consider monthly observations where $s = 12$, but the application to other values of s is straightforward. Box and Jenkins (1970) have generalized the ARIMA model to deal with seasonality and define a general multiplicative seasonal model in the form (using their notation)

$$\phi_p (B)\Phi_P(B^{12})W_t = \theta_q (B)\Theta_Q (B^{12})a_t \qquad (4.16)$$

where B denotes the backward shift operator, ϕ_p, Φ_P, θ_q, Θ_Q are polynomials of order p, P, q, Q respectively, and $\{a_t\}$ rather than $\{Z_t\}$ is the Box–Jenkins notation for a purely random process with mean zero and variance σ_a^2. The variables $\{W_t\}$ are formed from the original series $\{X_t\}$ by differencing to remove both trend and seasonality by

$$W_t = \nabla^d \nabla_{12}^D X_t$$

where for example if $d = D = 1$, then

$$W_t = \nabla \nabla_{12} X_t = \nabla_{12} X_t - \nabla_{12} X_{t-1}$$
$$= (X_t - X_{t-12}) - (X_{t-1} - X_{t-13}).$$

The values of the integer d and D do not usually need to exceed one. Also in equation (4.16) note that the shift operator B^{12} is such that

$$B^{12} W_t = W_{t-12}.$$

As an example of a seasonal model, suppose that $p = Q = 1$ and $P = q = 0$. Then (4.16) can be written

$$(1 - \alpha B) W_t = (1 + \theta B^{12}) a_t$$

so that $W_t = \alpha W_{t-1} + a_t + \theta a_{t-12}$.

If for example $W_t = \nabla_{12} X_t$, then we find

$$X_t = X_{t-12} + \alpha(X_{t-1} - X_{t-13}) + a_t + \theta a_{t-12}.$$

The reader should compare the seasonal model (4.16) with equations (3.5a) and (3.6).

When fitting a seasonal model to data, the first task is to assess values for p, P, q and Q by looking at the sample ac.f. and other indicators. Then the model parameters need to be estimated. This is discussed briefly on p. 91 and in full by Box and Jenkins (1970, Chapter 9).

75

4.7 Residual analysis

When a model has been fitted to a time series, it is advisable to check that the model really does provide an adequate description of the data. The usual way of doing this is to examine the *residuals*, which are the differences between the observations and the fitted values. For example with a first order AR model (equation 3.4), the true errors are given by

$$z_t = x_t - \alpha x_{t-1}$$

if we know the true value of the parameter α. When α is estimated by least squares, the fitted value is $\hat{\alpha} x_{t-1}$, and the residual is given by

$$\hat{z}_t = x_t - \hat{\alpha} x_{t-1}.$$

The analysis of residuals is discussed by Box and Jenkins (1970; Section 8.2). It is important to realize that the residuals are necessarily correlated even if the true errors are independent. From Section 4.1, we have that the correlogram of the true errors is such that each r_k is approximately normally distributed, mean 0, variance $1/N$, for reasonably large values of N. But the correlogram of the residuals has somewhat different properties. For example for a first-order AR process with $\alpha = 0.7$, the 95% confidence limits are at $\pm 1.3/\sqrt{N}$ for r_1, $\pm 1.7/\sqrt{N}$ for r_2, and $\pm 2/\sqrt{N}$ for values of r_k at higher lags. Thus for lags greater than 2 the confidence limits *are* the same as for the correlogram of the true errors, but for lags 1 and 2 they are not.

This type of problem was first investigated by J. Durbin and G. S. Watson in regard to multiple regression models applied to time-series data (Durbin and Watson, 1950, 1951, 1971). Suppose we have $(k + 1)$ time series of N observations with a dependent variable y and k 'independent' variables $x_1, \ldots, x_k$ and we fit the model

$$y_t = \beta_1 x_{1t} + \ldots + \beta_k x_{kt} + Z_t \qquad (t = 1, \ldots, N).$$

Having estimated the parameters β_i by least squares we want to see if the errors Z_t are independent. The residuals

$$\hat{z}_t = y_t - \hat{\beta}_1 x_{1t} - \ldots - \hat{\beta}_k x_{kt} \qquad (t = 1, \ldots, N)$$

are therefore calculated and the Durbin–Watson statistic is given by

$$d = \frac{\sum\limits_{t=2}^{N} (\hat{z}_t - \hat{z}_{t-1})^2}{\sum\limits_{t=1}^{N} \hat{z}_t^2} \qquad (4.17)$$

The distribution of d under the null hypothesis that the Z_t are independently distributed has been investigated and tables are available (e.g. Kendall, 1976) which give critical values of d. These depend on the number of independent variables in the model.

It is useful to note that d is related to the first autocorrelation coefficient of the residuals. From (4.17) we have

$$\sum\limits_{t=2}^{N} (\hat{z}_t - \hat{z}_{t-1})^2 \simeq 2 \sum\limits_{t=1}^{N} \hat{z}_t^2 - 2 \sum\limits_{t=2}^{N} \hat{z}_t \hat{z}_{t-1}$$

so that

$$d \simeq 2(1 - r_1)$$

where $r_1 = \Sigma \hat{z}_t \hat{z}_{t-1} / \Sigma \hat{z}_t^2$ is the first autocorrelation coefficient of the residuals (since mean residual is zero).

A test on d is therefore asymptotically equivalent to a test on r_1. A statistic similar to d can also be calculated to test for higher-order autocorrelation. Wallis (1972) considers the statistic

$$d_4 = \sum\limits_{t=5}^{N} (\hat{z}_t - \hat{z}_{t-4})^2 \Big/ \sum\limits_{t=1}^{N} \hat{z}_t^2$$

for testing for fourth order autocorrelation. Here we find $d_4 \simeq 2(1 - r_4)$.

Durbin and Watson make it clear that their statistic should only be used when the independent variables can be regarded as 'fixed' variables. Unfortunately the test has sometimes been used inappropriately for autoregressive processes, and as Nerlove and Wallis (1966) have pointed out, a non-significant value of d does not preclude the possibility that the errors are positively autocorrelated if the model contains autoregressive terms (or lagged endogenous variables in econometric jargon). However a significant value of d *does* indicate autocorrelated errors both for multiple regression and AR models. Durbin (1970) has considered the Durbin—Watson test for models containing autoregressive terms, but a more general approach to the problem is provided by Box and Pierce (1970), who consider the large sample properties of all the residual autocorrelation coefficients for any ARMA process. Their results, usefully summarized by Box and Jenkins (1970), show that $1/\sqrt{N}$ supplies an upper bound for the standard error of the r_k's calculated from the residuals. For low lags, particularly lag 1, the standard error of r_k may be substantially less than $1/\sqrt{N}$ but for higher lags $1/\sqrt{N}$ is a good approximation. Thus, if we calculate the residual correlogram in the usual way, values of r_k which lie outside the range $\pm 2/\sqrt{N}$ are certainly significantly different from zero. But for low lags, values of r_k inside the range $\pm 2/\sqrt{N}$ may still be significantly different from zero and need further investigation.

Box and Jenkins (1970, Section 8.2.2) also describe what they call a portmanteau lack-of-fit test. Instead of looking at the residual r_k's separately, one looks at the first K values of r_k together, by calculating

$$Q = N \sum_{k=1}^{K} r_k^2 \tag{4.18}$$

where N is the number of terms in the differenced series. If the fitted model is appropriate, then Q should be approximately distributed as χ^2 with $(K - p - q)$ degrees of freedom, where p, q are the number of AR and MA terms, respectively. Unfortunately, the χ^2 approximation does not work very well even for moderately large sample sizes such as 100. Various alternative statistics have been proposed (e.g. Ljung and Box, 1978) but the tests still seem to have rather poor power properties (Davies and Newbold, 1979) and in my experience rarely give a significant result. My own preference is just to look at the first few values of r_k, particularly at lag one and the first seasonal lag (if any), and see if any are significantly different from zero, although if just one value of r_k is 'just' significant there would not be enough evidence to reject the model.

4.8 General remarks on model-building

How do we set about finding a suitable model for a given time series? The answer depends on a number of factors including the properties of the series as assessed by a visual examination of the data, the number of observations available, and the way the model is to be used.

In Chapter 3 we introduced a general class of linear models called ARIMA models, which includes AR, MA and ARMA models as special cases. This useful class of processes provides a good fit to many different types of time series and should generally be considered when more than about 50 observations are available. Another general class of models is the trend and seasonal type of model introduced in Chapter 2. Various other models are also available.

In areas such as oceanography and electrical engineering, long stationary series often occur. If a parametric model is required, an ARMA model should be considered and can be fitted as outlined earlier in the chapter. The observed correlogram and the partial ac.f. are examined, the appropriate ARMA model identified, and the model parameters estimated by least squares. However, as we shall see in Chapters 6 and 7, we may be more interested in the frequency properties of the time-series in which case an ARMA model may not be very helpful.

In many other areas, such as economics and marketing, non-stationary series often occur and in addition may be fairly short. If more than 50 observations are available, Box and Jenkins (1970) advocate the fitting of ARIMA models by differencing the observed time series until it becomes stationary and then fitting an ARMA model to the differenced series. For seasonal series, the seasonal ARIMA model may be used. However it should be clearly recognized that when the variation of the systematic part of the time series (i.e. the trend and seasonality) is dominant, the effectiveness of the ARIMA model is mainly determined by the initial differencing operations and not by the subsequent fitting of an ARMA model to the differenced series, even though the latter operation is much more time-consuming. Thus the simple models discussed in Chapter 2 may be perfectly adequate for some time series with a pronounced trend and/or large seasonal effect. For example we might represent a series by a linear trend together with seasonal and error terms in the form

$$x_t = a + bt + s_t + \epsilon_t$$

where a, b are constants. Models of this type have the advantage of simplicity, of being easy to interpret, and of being fairly robust. In addition they can be used for short series where it is impossible to fit an ARIMA model. But they are restrictive with regard to the form of the trend and usually assume the errors to be independent which is sometimes not the case.

Exercises

1. Derive the least-squares estimates for a first-order AR process having mean μ.

2. Derive the least-squares normal equations for an AR(p) process, taking $\hat{\mu} = \bar{x}$, and compare with the Yule–Walker equations (equation 4.12).

3. For a second-order AR process, show that the (theoretical) partial autocorrelation coefficient of order 2 is given by

$$[\rho(2) - \rho(1)^2]/[1 - \rho(1)^2].$$

Compare with equation (4.11).

4. Suppose that the correlogram of a time series consisting of 100 observations has $r_1 = 0.31$, $r_2 = 0.37$, $r_3 = -0.05$, $r_4 = 0.06$, $r_5 = -0.21$, $r_6 = 0.11$, $r_7 = 0.08$, $r_8 = 0.05$, $r_9 = 0.12$, $r_{10} = -0.01$. Suggest an ARMA model which may be appropriate.

CHAPTER 5

Forecasting

5.1 Introduction

Forecasting the future values of an observed time series is an important problem in many areas, including economics, production planning, sales forecasting and stock control.

Suppose we have an observed time series, $x_1, x_2, \ldots, x_N$. Then the problem is to estimate x_{N+1}, or more generally x_{N+q}. The prediction of x_{N+q} made at time N of the value q steps ahead will be denoted by $\hat{x}(N, q)$. The integer q is called the *lead time*. There is, of course, no universally applicable forecasting procedure. Instead we must look for the procedure which is most appropriate for a given set of conditions.

The many types of forecasting procedure can be classified into three broad categories:

(a) Subjective   Forecasts can be made on a subjective basis using judgement, intuition, commercial knowledge and any other relevant information. Procedures of this type, such as the Delphi method, will not be described here. (See, for example, Chambers *et al.,* 1971; Keay, 1972).

(b) Univariate   Forecasts can be based entirely on past observations in a given time series, by fitting a model to the data and extrapolating. For example, forecasts of future sales of a product would be based entirely on past sales. Methods

82

of this type are sometimes called 'naive' or *'projection'* methods.

c) Multivariate   Forecasts can be made by taking observations on other variables into account. For example, sales may depend on stocks. Regression models are of this type, as are econometric models. The use of a 'leading indicator' also comes into this category. Multivariate models are sometimes called 'causal' or 'prediction' models.

In practice, a forecasting procedure may involve a *combination* of the above approaches. For example univariate forecasts are often computed, and then adjusted subjectively. In particular, marketing forecasts are often made by combining the projections or predictions developed statistically from past data, with the knowledge and insight of people deeply involved in the market.

Another type of combination is that proposed by Bates and Granger (1969) where two or more objective forecasts are computed and then a weighted average is calculated.

Before choosing a forecasting procedure, it is essential to consider how the forecast is to be used, what accuracy is required, how much money is available, how many items are to be forecast, how much data is available, and how far ahead forecasts are required. We will return to these questions in Section 5.4.

The forecasting procedure may depend somewhat on the required lead time. We will be mainly concerned with what is called *short-term* forecasting where the lead time is less than about nine months. For example in stock control, the lead time for which forecasts are required is the time between ordering an item and its delivery, which is usually a few weeks or months. Longer term forecasts are more often required in government and corporate planning and then it is generally advisable to use a fairly simple procedure. The longer the lead time, the more likely is the underlying model to change and hence the more unreliable the forecast.

Some forecasting procedures simply produce *point* forecasts. But these do not indicate the uncertainty associated with the estimation of future demand. Thus it is usually desirable to produce an *interval* forecast. Some procedures, such as Box–Jenkins, enable one to do this but others do not, in which case upper and lower limits may have to be assessed on a subjective basis.

5.2 Univariate procedures

This section introduces the many projection methods which are now available. Further details may be found in Granger and Newbold (1977), Montgomery and Johnson (1977) and Gilchrist (1976).

For all the procedures, the first step is to *plot* the data as much useful information can often be obtained from a visual examination of the data, and this may help to suggest an appropriate forecasting procedure.

5.2.1 Extrapolation of trend curves

For long-term forecasting it is often useful to fit a trend curve to successive yearly totals and extrapolate. This problem is discussed by Gregg *et al.* (1964) and Harrison and Pearce (1972). A variety of curves may be tried including polynomial, exponential and Gompertz curves. At least seven to ten years of historical data are required and Harrison and Pearce suggest that 'one should not make forecasts for a longer period ahead than about half the number of past years for which data are available'.

For seasonal data, the trend and seasonal variation may be estimated by the methods of Chapter 2, either by using moving averages or by fitting a trend curve. The fitted curve may then be extrapolated into the future. The method is simple, fairly crude, but is robust and cheap and is worth considering for long-term forecasting where it is unlikely to be worthwhile fitting a complicated model to past data as the model may change in the future.

A drawback to the use of trend curves is that there is no logical basis for choosing among the different curves except by goodness-of-fit. Unfortunately it is often the case that one can find several curves which fit a given set of data almost equally well, but which when projected forward, give widely different forecasts.

5.2.2 Exponential smoothing This forecasting procedure, first suggested by C. C. Holt in about 1958, should only be used in its basic form for non-seasonal time series showing no trend. Of course many time series which arise in practice do contain a trend or seasonal pattern, but these effects can be measured and removed to produce a stationary series. Thus it turns out that adaptations of exponential smoothing can be used for most types of time-series and so the following remarks provide a useful basis for a study of the methods due to Brown (1963), Holt–Winters, Harrison (1965), and Box and Jenkins (1970).

Given a stationary, non-seasonal, time series, $x_1, x_2, \ldots, x_N$, it is natural to take as an estimate of x_{N+1} a weighted sum of the past observations.

$$\hat{x}(N, 1) = c_0 x_N + c_1 x_{N-1} + c_2 x_{N-2} + \ldots \qquad (5.1)$$

where the $\{c_i\}$ are weights. It seems sensible to give more weight to recent observations and less weight to observations further in the past. An intuitively appealing set of weights are geometric weights which decrease by a constant ratio. In order that the weights sum to one, we take

$$c_i = \alpha(1 - \alpha)^i \qquad (i = 0, 1, \ldots)$$

where α is a constant such that $0 < \alpha < 1$. Then (5.1) becomes

$$\hat{x}(N, 1) = \alpha x_N + \alpha(1 - \alpha)x_{N-1} + \alpha(1 - \alpha)^2 x_{N-2} + \ldots. \qquad (5.2)$$

Strictly speaking, equation (5.2) implies an infinite number of past observations, but in practice there will only be a finite number. So equation (5.2) is customarily rewritten as

$$\hat{x}(N, 1) = \alpha x_N + (1 - \alpha)[\alpha x_{N-1} + \alpha(1 - \alpha)x_{N-2} + \dots]$$
$$= \alpha x_N + (1 - \alpha)\hat{x}(N - 1, 1). \qquad (5.3$$

If we set $\hat{x}(1, 1) = x_1$, then equation (5.3) can be used recursively to compute forecasts. Equation (5.3) also reduces the amount of arithmetic involved since forecasts can easily be updated using only the latest observation and the previous forecast.

The procedure defined by equation (5.3) is called exponential smoothing and has proved very useful in many forecasting situations. The adjective 'exponential' arises from the fact that the geometric weights lie on an exponential curve, but the procedure could equally well be called geometric smoothing.

Equation (5.3) is sometimes rewritten in the form

$$\hat{x}(N, 1) = \alpha[x_N - \hat{x}(N - 1, 1)] + \hat{x}(N - 1, 1)$$
$$= \alpha e_N + \hat{x}(N - 1, 1) \qquad (5.4$$

where $\quad e_N = x_N - \hat{x}(N - 1, 1)$

$\qquad \qquad$ = prediction error at time N.

It can be shown that exponential smoothing is optimal if the underlying model for the time series is given by

$$X_t = \mu + \alpha \sum_{j < t} Z_j + Z_t \qquad (5.5$$

This infinite moving average process is non-stationary, but the first differences $(X_{t+1} - X_t)$ form a first-order moving average process so that X_t is an ARIMA $(0, 1, 1)$ process.

The value of the smoothing constant, α, depends on the properties of the given time series. Values between 0.1 and 0.3 are commonly used and produce a forecast which depends on a large number of past observations. Values close to one are used rather less often and give forecasts which

epend much more on recent observations. When $\alpha = 1$, the
orecast is equal to the most recent observation.

The value of α may be estimated from past data by a
milar procedure to that used for estimating the parameters
f a moving average process. The sum of squared prediction
rrors is computed for different values of α and the value is
hosen which minimizes the sum of squares. With a given
alue of α, calculate

$$\hat{x}(1, 1) = x_1$$
$$e_2 = x_2 - \hat{x}(1, 1)$$
$$\hat{x}(2, 1) = \alpha e_2 + \hat{x}(1, 1)$$
$$e_3 = x_3 - \hat{x}(2, 1)$$
$$\dots \dots \dots \dots \dots$$
$$e_N = x_N - \hat{x}(N - 1, 1)$$

nd compute $\sum\limits_{i=2}^{N} e_i^2$. Repeat this procedure for other values
f α between 0 and 1, say in steps of 0.1, and select the value
hich minimizes Σe_i^2, (or omit the first few e's and minimize
$\sum\limits_{c}^{N} e_i^2$). Usually the sum of squares surface is quite flat near
he minimum and so the choice of α is not critical.

.2.3 Holt—Winters forecasting procedure   Exponential
moothing may easily be generalized to deal with time series
ontaining trend and seasonal variation. The resulting
rocedure is usually referred to as the Holt—Winters
rocedure and is described by Winters (1960) and Chatfield
1978a). Trend and seasonal terms are introduced which are
lso updated by exponential smoothing.

Suppose the observations are monthly. Let m_t denote the
stimated current mean in month t, r_t denote the estimated

trend term in month t (i.e. the expected increase or decrease per month in the current mean), and s_t denote the estimated seasonal factor appropriate to month t. Then, as each new observation becomes available, all three terms are updated. If the seasonal variation is multiplicative (model B of Section 2.4), the updating equations are:

$$m_t = \alpha x_t / s_{t-12} + (1 - \alpha)(m_{t-1} + r_{t-1})$$
$$s_t = \beta x_t / m_t + (1 - \beta)s_{t-12}$$
$$r_t = \gamma(m_t - m_{t-1}) + (1 - \gamma)r_{t-1}.$$

where x_t is the latest observation and α, β, γ are constants such that $0 < \alpha, \beta, \gamma < 1$. The forecasts from time t are then

$$\hat{x}(t, h) = (m_t + hr_t)s_{t-12+h} \qquad (h = 1, 2, \ldots, 12)$$

If the seasonal variation is additive, the updating equations are:

$$m_t = \alpha(x_t - s_{t-12}) + (1 - \alpha)(m_{t-1} + r_{t-1})$$
$$s_t = \beta(x_t - m_t) + (1 - \beta)s_{t-12}$$
$$r_t = \gamma(m_t - m_{t-1}) + (1 - \gamma)r_{t-1}.$$

A graph of the data should be examined to see if an additive or multiplicative seasonal effect is the more appropriate. The method may 'blow up' if the wrong model is used. If the seasonal period does not cover 12 observations, then the equations need to be modified in an obvious way. Starting values for m_t, r_t and s_t may be estimated in a fairly crude way from the first two years data, by taking

$$m_1 = \sum_{t=1}^{12} x_t/12, r_1 = \text{(mean of 2nd year} - \text{mean of 1st year)}/12$$

and $s_1, \ldots, s_{12}$ to be the average seasonal effects in the first two years when the different months are compared with the yearly means. The three smoothing constants, α, β, and γ, are chosen as in Section 5.2.2 except that on this occasion the quantity to be minimized will be $\sum\limits_{i=25}^{N} e_i^2$.

If the seasonal smoothing constant, β, is close to zero, then the updating equations for s_t may need modification. The updating technique for exponential smoothing effectively assumes an infinite amount of past data. But if say six years monthly data is available, then each of the monthly factors will only be revised five times and inordinate weight may be given to the starting value. For example if $\beta = 0.1$ and the starting value for month 1 is s_1, then

$$s_{13} = 0.1x_{13}/m_{13} + 0.9s_1$$

and we eventually find

$$s_{61} = 0.1x_{61}/m_{61} + 0.09x_{49}/m_{49} + \ldots + 0.9^5 s_1.$$

This undesirable effect can be removed by using higher values for β in the first few years. One strategy is to take $\beta = 1/n$ in year n up to a maximum of about $n = 8$. Apart from the choice between multiplicative and additive seasonality, the Holt–Winters procedure can be made fully automatic in the sense that a single computer program can be written so as to produce forecasts for any given time series without human intervention, and the technique is widely used in industry.

5.2.4 Box–Jenkins forecasting procedure   This forecasting procedure has been developed by G. E. P. Box and G. M. Jenkins over the last 20 years. This section only gives a brief broad outline of the procedure. Further details may be found in Box and Jenkins (1970) although the beginner may find it easier to read Box and Jenkins (1968), Nelson (1973), Chatfield and Prothero (1973a), Granger and Newbold (1977) or Jenkins (1979). Recent improvements in model identification techniques are discussed in Chapter 10.

The procedure is based on fitting an autoregressive integrated moving average (ARIMA) model to a given set of data and then taking conditional expectations. The main stages in setting up a Box—Jenkins forecasting model are as follows:

(a) Model identification   Examine the data to see which member of the class of ARIMA processes appears to be most appropriate.

(b) Estimation   Estimate the parameters of the chosen model by least squares as described in Chapter 4.

(c) Diagnostic checking   Examine the residuals from the fitted model to see if it is adequate.

(d) Consider alternative models if necessary   If the first model appears to be inadequate for some reason, then other ARIMA models may be tried until a satisfactory model is found.

Now AR, MA and mixed models have been around for many years and are associated in particular with G. U. Yule and H. Wold. The major contribution of Box and Jenkins has been to provide a general *strategy* for time-series forecasting. By stressing differencing, they enable one to set up models for non-stationary series and the general class of ARIMA models is often referred to as the Box—Jenkins class of models. They have also provided general estimation and diagnostic checking procedures.

The first step in the Box—Jenkins procedure is to difference the data until it is stationary. This is achieved by examining the correlograms of various differenced series until one is found which comes down to zero 'fairly quickly' and from which any seasonal effect has been largely removed. For non-seasonal data, first-order differencing is usually sufficient. For seasonal data of period 12, the operator $\nabla \nabla_{12}$ is often used if the seasonal effect is additive, while the operator ∇_{12}^2 may be used if the seasonal effect is multiplicative. (See exercise 1.) Sometimes the operator ∇_{12} by itself will be sufficient. Over-differencing should be avoided. For quarterly data, the operator ∇_4 may be used, and so on.

The differenced series will be denoted by $\{w_t; t = 1, \ldots, N-c\}$ where c terms are 'lost' by differencing. For example, if the operator $\nabla\nabla_{12}$ is used, then $c = 13$.

If the data is non-seasonal, an ARMA model can now be fitted to $\{w_t\}$ as described in Chapter 4. If the data is seasonal, then the general seasonal ARMA model defined in equation 4.16 may be fitted as follows. 'Reasonable' values of p, P, q, Q are selected by examining the correlogram of the differenced series $\{w_t\}$. Values of p and q are selected as outlined in Chapter 4 by examining the first few values of r_k. Values of P and Q are selected by examining the values of r_k at $k = 12, 24 \ldots$ (where the seasonal period is 12). If for example r_{12} is 'large' but r_{24} is 'small', this suggests one seasonal moving average term so we would take $P = 0$, $Q = 1$.

Having tentatively identified what appears to be a reasonable seasonal ARMA model, least-squares estimates of the model parameters may be obtained by minimizing the residual sum of squares in a similar way to that proposed for ordinary ARMA models. In the case of seasonal series, it is advisable to estimate initial values of a_t and w_t by back-forecasting (or backcasting) rather than set them equal to zero. This procedure is described by Box and Jenkins (1970, Section 9.2.4). In fact if the model contains a seasonal moving average parameter which is close to one, several cycles of forward and backward iteration may be needed.

For both seasonal and non-seasonal data, the adequacy of the fitted model should be checked by what Box and Jenkins call 'diagnostic checking'. This essentially consists of examining the residuals from the fitted model to see if there is any evidence of non-randomness. The correlogram of the residuals is calculated and we can then see how many

coefficients are significantly different from zero and whether any further terms are indicated for the ARIMA model.
If the fitted model appears to be inadequate, then alternative ARIMA models may be tried until a satisfactory one is found.

When a satisfactory model is found, forecasts may readily be computed. Given data up to time N these forecasts will involve the observations and fitted residuals up to and including time N. General methods of finding forecast formulae are described fully by Box and Jenkins (1970). They rely on the fact that the best (i.e. least-squares) forecast of all future residuals $(a_{N+1}, a_{N+2}, \ldots)$ is simply zero. The forecasts can often be found using common sense. For example, consider the seasonal ARIMA model example given on p. 75 where

$$X_t = X_{t-12} + \alpha(X_{t-1} - X_{t-13}) + a_t + \theta a_{t-12}.$$

The forecast of x_{N+1} is 'obviously' given by

$$\hat{x}(N,1) = x_{N-11} + \hat{\alpha}(x_N - x_{N-12}) + \hat{\theta}\, \hat{a}_{N-11}$$

where $\hat{\alpha}$, $\hat{\theta}$ are estimates of α, θ and $\hat{a}_{N-11}$ is the estimated residual at time $(N-11)$. The forecast of x_{N+2} will involve the (unknown as yet) value of x_{N+1} and so we insert the forecast value instead (see Exercise 4).

In contrast to the Holt–Winters forecasting procedure which assumes a simple family of models, the Box–Jenkins procedure is not fully automatic. A model is chosen from the large class of ARIMA models according to the properties of the particular time-series being analysed. Thus while the procedure is more versatile than its competitors it is also more complicated and expensive, and considerable experience is required to identify an appropriate ARIMA model. One difficulty that sometimes arises is that several ARIMA models may be found which fit the data equally well. Such models will often be almost equivalent and will therefore yield nearly identical forecasts, but this is not always the case. Another drawback to the method is that at least 50, and preferably 100, observations are needed for it to have a chance of success. We will make further comments on the applicability of the method in Section 5.4.

5.2.5 Stepwise autoregression

Granger and Newbold (1977, Section 5.4) have described a procedure called step-wise autoregression which can be regarded as a subset of the Box–Jenkins procedure and which has the advantage of being fully automatic. The method relies on the fact that AR models are much easier to fit than MA or ARMA models even though an AR model may require extra parameters to get as good a representation of the data.

First differences of the data are taken to allow for non-stationarity in the mean. Then a maximum possible lag, say p, is chosen. The best autoregressive model with just one lagged variable is then found:—

$$W_t = \mu + \alpha_s^{(1)} W_{t-s} + e_t^{(1)} \qquad 1 \leqslant s \leqslant p.$$

where $W_t = X_t - X_{t-1}$, and $\alpha_s^{(1)}$ is the autoregression coefficient. Then the best autoregressive model with 2, 3, ... lagged variables is found. The procedure is terminated when the reduction in the sum of squared residuals at the jth stage is less than some pre-assigned quantity.

Thus an integrated autoregressive model is fitted which is a special case of the Box–Jenkins ARIMA class.

Newbold and Granger suggest choosing $p = 13$ for quarterly data and $p = 25$ for monthly data.

5.2.6 Other methods

Several other forecasting procedures have been proposed. Brown (1963) has suggested a technique called *general exponential smoothing* which consists of fitting polynomial, sinusoidal or exponential functions to the data and finding appropriate updating formula. One special case of this is double exponential smoothing which is applicable to series containing a linear trend. Note that Brown suggests fitting by *discounted* least squares, in which more weight is given to recent observations. Generally speaking, Brown's method compares unfavourably with other univariate forecasting procedures, especially on seasonal data (see Reid, 1975).

Harrison (1965) has proposed a modification of seasonal exponential smoothing which consists essentially of performing a Fourier analysis of the seasonal factors and replacing them by smoothed factors. Harrison and Stevens (1976) have proposed a technique, called *Bayesian forecasting* which depends on a model called 'the dynamic linear model' whose parameters are updated by a technique called Kalman filtering (e.g. see Astrom, 1970; Young, 1974). Bayesian forecasting also allows one to specify a range of models, rather than a single model, with associated probabilities which are updated after each new observation. The method is claimed to be able to cope with outliers, step changes in mean and trend, and changes in the process parameters. This author has no practical experience with the method, but see Taylor and Thomas (1982).

Another proposed technique called *adaptive filtering* (Makridakis and Wheelwright, 1977) appears to be technically unsound (Chatfield, 1978b).

Various other projection methods can also be found in the literature but we do not have space to cover them all here.

5.3 Multivariate procedures

This section provides a brief introduction to multivariate forecasting procedures.

5.3.1 Multiple regression This approach uses the multiple linear regression model where the variable of interest (say y) is linearly related to present and past values of other variables (say $x_1, \ldots, x_p$) and also possibly to past values of y (these are autoregressive terms).

A description of multiple regression can be found in many statistics textbooks (e.g. Anderson, 1971) and will not be repeated here. With the general availability of multiple regression computer programs it is computationally easy (perhaps too easy!) to fit a multiple regression model and use it for planning or forecasting.

Multiple regression models sometimes work well, particularly in a marketing context. But there are several dangers in the method which need to be appreciated. Firstly the ready availability of computer programs has resulted in a tendency to put more and more explanatory variables into the model with dubious results. The resulting model may indeed appear to give a good fit to the available data. By including say 20 explanatory variables one may achieve a multiple correlation coefficient, R^2, as high as 0.995, but this good fit may be spurious and does not necessarily mean that the model will give good forecasts. A more sensible number of explanatory variables is a maximum of 6 or 7, and it is advisable to fit the model to part of the available data and then check the model by forecasting the remainder of the data.

In marketing applications the so-called 'independent' variables are usually not independent at all, and if some of them are highly correlated there may be singularity problems. It is therefore advisable to look at the correlation matrix of the 'independent' variables *before* carrying out a multiple regression so that, if necessary, some variables can be excluded. It is unnecessary for the explanatory variables to be completely independent, but large correlations should be avoided.

Another difficulty arising in multiple regression is that

some crucial explanatory variables may have been held more or less constant in the past and it is then impossible to assess their effect and include them in the model in a quantitative way. For example a company may be considering increasing its advertising expenditure and would like to construct a model which would predict the effect on sales. But if advertising has been held relatively constant in the past, then it will be impossible to estimate the effect of advertising, and yet a model which excludes advertising may be useless if advertising expenditure is changed.

But perhaps the most important problem in multiple regression forecasting concerns the structure of the error terms. It is often assumed that these are an independent white noise sequence, but such an assumption is sometimes not appropriate and may lead to disappointing results (Granger and Newbold, 1974), or to an inappropriate model (Box and Newbold, 1971). Having fitted a multiple regression model, one should check the residuals for autocorrelation as described in Section 4.7. If the residuals are autocorrelated, one can try fitting a multiple regression model with autocorrelated errors by a method, due to D. Cochrane and G. H. Orcutt, which is described by Kendall (1976; Section 12.9). Alternatively one can try the Box—Jenkins approach.

In summary, I am inclined to agree with Brown (1963, p. 77) that the use of multiple regression models can be very dangerous except in certain special cases where one has a definite reason why one series should be related to another. An example of such a situation, involving just one explanatory variable, is given in example 3 of Section 5.5. Another example is given by Bhattacharyya (1974).

5.3.2 Econometric models Econometric models (e.g. Harvey, 1981) often assume that an economic system can be described, not by a single equation, but by a set of simultaneous equations. For example, not only do wage rates depend on prices but also prices depend on wage rates. Economists distinguish between exogenous variables, which affect the system but are not themselves affected, and endogenous variables which interact with each other. The simultaneous equation system involving k dependent (endogenous) variables, $\{Y_i\}$, and g pre-determined (exogenous) variables, $\{X_i\}$, may be written

$$Y_i = f_i(Y_1, \ldots, Y_{i-1}, Y_{i+1}, \ldots, Y_k, X_1, \ldots, X_g)$$
$$+ \text{error} \qquad (i = 1, 2, \ldots, k).$$

Some of the exogenous variables may be lagged values of the Y_i. These equations, often called the structural form of the system, can be solved to give what is called the reduced form of the system, namely

$$Y_i = F_i(X_1, \ldots, X_g) + \text{error} \qquad (i = 1, 2, \ldots k).$$

The functions $\{f_i\}$, $\{F_i\}$ are usually assumed to be linear and the form of the equations is often specified *a priori* from economic theory. The uncontrolled nature of much economic data (e.g. see Pierce, 1977) makes it difficult to construct econometric models empirically and we shall say no more about it here. A simple example with just two equations is given by Leuthold *et. al* (1970). An example with two equations in animal population dynamics is given by Chan and Wallis (1978).

5.3.3 Box—Jenkins method   Box and Jenkins (1968 and 1970) consider multivariate forecasting as well as univariate forecasting. They propose a class of models, called transfer function models (see Chapter 9), which is a natural extension of the ARIMA class. They concentrate on the case of describing the relationship between one 'output' variable (e.g. sales) and one 'input' variable (e.g. advertising expenditure), though the method can be generalized to cover several input variables. They show how to identify an appropriate model and then make predictions. This method will be introduced in Section 9.4.2.

5.4 A comparison of forecasting procedures

Many factors need to be considered when choosing the forecasting procedure which is most appropriate for a given set of conditions. These factors include the purpose of the forecast, the degree of accuracy required, and the amount of money available. No clear-cut rules can be given but the following remarks are intended as a general guide.

The most important consideration is how the forecast is to be used. Many forecasts are used for planning purposes, for example in production planning or stock control. Alternatively, a forecast may act as a 'norm' or yardstick. Sometimes more than one forecast may be required. For example suppose that a company wants to carry out an advertising campaign so as (hopefully) to increase sales. Then they may require a 'norm' forecast based only on past sales so that any increase in sales may be assessed, and may also require a planning forecast based on all available information which will try to forecast the effect of the sales campaign so as to ensure that ample product is available.

Other considerations in choosing a forecasting procedure include the required accuracy of the forecasts, how much account should be taken of changes in other factors, and how important the past is thought to be in determining the future. If the model is thought likely to change, then it is not worth spending too much time fitting a complicated model to past data. Whatever forecasting method is adopted, one is still extrapolating from the past into the future, with all the inherent dangers. Clearly, one should always be prepared to modify forecasts in the light of any other information.

As they do not take other variables into account, univariate forecasts are mainly intended to act as a 'norm'. If forecasts are required for planning or decision-making, then ideally one should set up a multivariate model. However, multivariate models are relatively difficult and expensive to build, particularly econometric and transfer function models. It is also fair to say that econometric models have often turned out to have poor predictive powers. For example, Naylor *et al.* (1972) show that univariate Box—Jenkins forecasts are better than forecasts obtained using the Wharton econometric model. Thus although one would expect multivariate models to give better forecasts than univariate models, this is not necessarily so.

Of the univariate procedures, Reid (1975) and Newbold and Granger (1974) show that the Box—Jenkins method tends to give more accurate forecasts than simple automatic procedures, provided the former is carried out by an 'expert', while Chatfield (1978a) shows that automatic procedures can often be improved in particular cases by common-sense modifications. Holt—Winters, Harrison and step-wise autoregression are roughly equal in accuracy but better than Brown's method. Newbold and Granger (1974) found that a combination of Holt- Winters and step-wise autoregression (which are both automatic procedures) gave results which were almost as accurate as those given by the Box—Jenkins approach.

The choice of forecasting procedure also depends on the number of items which have to be forecast. If a large number of items are involved, say several hundred, then for production planning and stock control purposes it may be useful to set up a computer forecasting system to allow routine forecasts to be made without human intervention. Then a univariate procedure is appropriate. The high initial cost of the Box—Jenkins procedure will make it unsuitable in this sort of situation. Instead, one will probably prefer a fully automatic procedure such as Holt–Winters or step-wise autoregression.

For a company that manufactures only a few items, Coutie *et al.* (1964) suggest that the salesman concerned can probably produce the best forecasts by subjective methods. This is sometimes frowned upon in academic circles as being 'unscientific', but there is some evidence to suggest that subjective methods may often be superior though the evidence is not all one way. Gearing (1971) gets better results using a rule-of-thumb technique than with an adaptive smoothing technique. Wagle *et al.* (1968) found several products for which hand forecasts are markedly better than objective statistical forecasts. Armstrong and Grohman (1972) discuss an example where econometric forecasts are better than univariate forecasts, which in turn are better than subjective forecasts. Myers (1971) says that forecasting should never be entirely objective and recommends extrapolating by eye and then estimating the effects of external factors on a subjective basis. My own preference in sales forecasting would generally be to compute univariate forecasts using a simple procedure such as Holt–Winters and then, if desired, to adjust the forecasts subjectively.

Finally let us attempt an assessment of the Box–Jenkins procedure (see also Chatfield, 1977, Section 5; and 1978a). Most univariate forecasting procedures are fully automatic in that only one set of forecasts can be produced from a given set of data. However, the Box–Jenkins procedure involves a subjective element which allows one to choose from a wide class of models. This greater versatility is both the strength and weakness of the Box–Jenkins approach. The advantage of being able to choose from a wide class of models rather than being restricted to one particular model is clear. At the same time, the subjective assessment involved in choosing a model means that considerable experience is required in interpreting sample autocorrelation functions. In addition the procedure is expensive in terms of both computing and staff time.

In many time-series situations, such as analysing macro-economic series, a large expenditure of time and effort can certainly be justified and then the method is well worth considering, provided that sufficient expertise is available. However for routine sales forecasting we take the view that the method is usually inappropriate for several reasons. Firstly, when variation is dominated by trend and seasonality, the effectiveness of the ARIMA model is mainly determined by the differencing procedure and some other way of measuring trend and seasonality may be more appropriate. Secondly, many firms do not have sufficient expertise available to employ the method effectively: Thirdly, the method may be too expensive: Fourthly, a fully automatic procedure is often appropriate: Fifthly, a manager is more likely to accept statistical forecasts if he understands the way they have been derived. Most managers understand the ideas of trend and seasonality, but would be completely baffled by ARIMA models. As G. J. A. Stern has pointed out, to introduce Box–Jenkins methods in some firms would require translating managers from the statistical stone age to the statistical space age in one leap, which is unrealistic. Nevertheless the ideas behind the Box–Jenkins method are undoubtedly of interest to any statistician interested in forecasting. With suitable development, they should become a useful tool in the time-series analyst's kit. For example Bhattacharyya (1974) gives an example combining regression and Box–Jenkins.

5.5 Some examples

In this section we discuss 3 sets of data, to illustrate some of the problems which arise in real forecasting situations.

Example 1   Fig. 5.1 shows the (coded) sales of a certain company in successive quarters over 6 years. Suppose that a

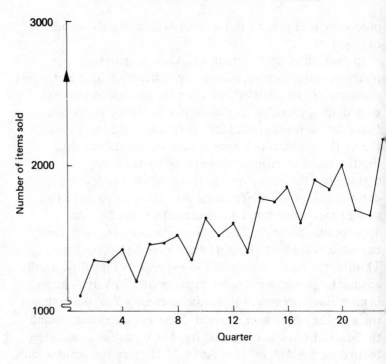

*Figure 5.1   Sales of a certain company in successive three-month
periods.*

univariate forecast is required for the next four quarters.
What method is appropriate? This series demonstrates the
importance of plotting a time series and making a visual
examination before deciding on the appropriate forecasting
procedure. It is evident from Fig. 5.1 that there is an
increasing trend and a pronounced seasonal effect with
observations 1, 5, 9, 13, . . ., relatively low. The series is too
short to use the Box–Jenkins method. Instead a suitable
forecasting procedure might appear to be that of
Holt–Winters. But closer examination of Fig. 5.1 reveals that
the observation in quarter 22 is unusually low while the
following observation seems somewhat high. If we were to
apply Holt–Winters with no modification, these unusual

ɔbservations would have a marked effect on the forecasts. We must therefore first decide if they indicate a permanent change in the seasonal pattern, in which case earlier ɔbservations will have little relevance for forecasting purposes, or if they were caused by some unusual phenomenon such as a strike. In the latter case, some adjustment of the observations in quarters 22 and 23 needs to be made before forecasts are computed.

Example 2   Fig. 5.2 shows some telephone data analysed by Tomasek (1972) using the Box—Jenkins method. He developed the model

$$(1 - 0.84B)(1 - B^{12})(X_t - 132)$$
$$= (1 - 0.60B)(1 + 0.37B^{12})a_t,$$

which when fitted to all the data explained 99.4% of the total variation about the mean. [i.e. the total corrected sum of squares, $\Sigma(x_t - \bar{x})^2$]. On the basis of this good fit, Tomasek recommended the use of the Box—Jenkins method for forecasting.

However, it is not at all clear that this is a sensible recommendation. Looking at Fig. 5.2, we see that the series has an unusually high regular seasonal pattern. In fact 97% of the variation about the mean is explained by a linear trend and constant seasonal pattern. As we remarked in Section 4.8, when the variation due to trend and seasonality is dominant, the effectiveness of the ARIMA model is mainly determined by the initial differencing operations and not by the time-consuming ARMA model fitting to the differenced series (Akaike, 1973). For such regular data, nearly any forecasting method will give good results. For example the Holt—Winters method explains 98.9% of the variation and it is rather doubtful if the extra expense of the Box—Jenkins method can be justified by increasing the explained variation from 98.9% to 99.4%.

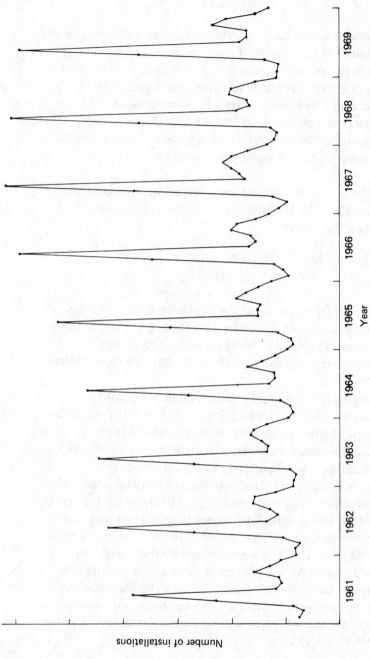

Number of installations

Year

1961   1962   1963   1964   1965   1966   1967   1968   1969

Figure 5.2   Tomasek's telephone data. Numbers of newly installed telephones in successive months. (Tomasek does not provide a vertical scale.)

Figure 5.3 Sales of company C in successive three-month periods.

Example 3   Fig. 5.3 shows quarterly sales data for company C over 12 successive years. Although there is some evidence of a seasonal pattern it is not particularly regular. D. L. Prothero tried two univariate procedures on this data, namely Holt—Winters and Box—Jenkins. The Box—Jenkins model fitted was

$$\nabla\nabla_4 X_t = (1 - 0.2B)(1 - 0.8B^4)a_t$$

The mean absolute forecast errors up to four quarters ahead were calculated as follows.

No. of quarters ahead	Holt—Winters	Box—Jenkins
1	2543	2454
2	2864	2988
3	3174	3416
4	3732	4143

Thus although Box—Jenkins is 3% better one step ahead, it is up to 10% worse four steps ahead.

This result is typical in the sense that Box—Jenkins tends to do less well as the lead time increases. But the result is not typical in that for series as irregular as that shown in Fig. 5.3, Box—Jenkins will sometimes do considerably better than other methods and, if sufficient money and expertise are available, then the Box—Jenkins method is worth a try for data of this type.

This data also illustrates the possible advantages of multivariate forecasting. It was found that if detrended, deseasonalized sales are linearly regressed on detrended, deseasonalized stocks two quarters before, the mean absolute forecast error one step ahead was 1900 which is considerably better than either of the univariate procedures we tried. In other words stocks are a leading indicator for sales. This illustrates another general point that if one wants to put in a lot of effort to get a good forecast, it may well be better to try a multivariate procedure such as multiple regression rather than a complicated univariate procedure such as Box–Jenkins, although this does not always follow (Naylor *et al.*, 1972).

†5.6 Prediction theory

Over the last thirty years or so, a general theory of linear prediction has been developed by Kolmogorov, Wiener (1949), Grenander and Rosenblatt (1957), Yaglom (1962), Whittle (1963) and Bartlett (1966) among others. All these authors avoid the use of the word 'forecasting' although most of the univariate methods considered in Section 5.2 are in the general class of linear predictors. The theory of linear prediction has applications in control and communications engineering and is of considerable theoretical interest, but readers who wish to tackle the sort of forecasting problem we have been considering earlier in this chapter will find this literature less accessible than the other references. Here we will only give a brief introduction.

Two types of problem are often distinguished. In the first type of problem we have data up to time T, $\{x_T, x_{T-1} \cdots\}$, and wish to predict the value of x_{T+m}. One approach is to use the predictor

$$\hat{x}_{T+m} = \Sigma c_j x_{T-j}$$

which is a linear function of the available data. The weights $\{c_j\}$ are chosen so as to minimize the expected mean square prediction error, $E(x_{T+m} - \hat{x}_{T+m})^2$. This is often called the *prediction* problem (e.g. Cox and Miller, 1968), while Yaglom (1962) refers to it as the *extrapolation* problem, and Whittle (1963) calls it *pure prediction*. As an example of the sort of result which has been obtained, Wiener (1949) has considered the problem of evaluating the weights $\{c_j\}$, so as to find the best linear predictor, when the ac.f. of the series $\{x_t\}$ is *known* and when the entire past of the sequence $\{x_t\}$ is known. It is interesting to compare this sort of approach

†This section may be omitted at first reading.

with the forecasting techniques proposed earlier in this chapter. The Box—Jenkins approach, for example, also employs a linear predictor which will be optimal for a particular ARIMA process. But whereas Wiener says little about estimation, Box and Jenkins show how to find a linear predictor when the ac.f. has to be estimated.

The second type of problem arises when the process of interest, $s(t)$, called the *signal*, is contaminated by *noise*, $n(t)$, and we actually observe the process

$$y(t) = s(t) + n(t).$$

In some situations the noise is simply measurement error. In engineering applications the noise is an interference process of some kind. The problem now is to separate the signal from the noise. Given measurements on $y(t)$ up to time T we may want to reconstruct the signal up to time T or alternatively make a prediction of $s(T + \tau)$. The problem of reconstructing the signal is often called *smoothing* or *filtering*. The problem of predicting the signal is also often called *filtering* (Yaglom, 1962; Cox and Miller, 1968), but is sometimes called *prediction* (Astrom, 1970). It is often assumed that the signal and noise processes are uncorrelated and that $s(t)$ and $n(t)$ have known ac.f's.

It is clear that both the above types of problem are closely related to the control problem because, if we can predict how a process will behave, then we can adjust the process so that the achieved values are, in some sense, as close as possible to the target value. Further remarks on control theory must await a study of linear systems.

Exercises

1. (a) If $x_t = (a + bt)s_t + \epsilon_t$ where a, b are constants, $s_t = s_{t-12}$, and ϵ_t is a white noise process, show that $\nabla_{12}^2 x_t$ is a stationary process.

   (b) If $x_t = a + bt + s_t + \epsilon_t$, show that $\nabla\nabla_{12} x_t$ is a stationary process.

2. For the AR(1) model given by

$$X_t = \alpha X_{t-1} + Z_t$$

show that the least-squares k-steps ahead forecast of X_{N+k} given data up to time N is given by

$$\hat{x}(N, k) = \alpha^k x_N \qquad (k = 1, 2, \ldots).$$

In practice one would substitute an estimate of α into the above forecast formula.

If the AR(1) model is given by

$$X_t - \mu = \alpha(X_{t-1} - \mu) + Z_t$$

show that $\hat{x}(N, k) = \mu + \alpha^k (x_N - \mu) \qquad (k = 1, 2, \ldots).$

3. For the MA(1) model given by

$$X_t = Z_t + \theta Z_{t-1}$$

show that

$$\hat{x}(N, 1) = \theta z_N$$

and

$$\hat{x}(N, k) = 0 \qquad (k = 2, 3, \ldots).$$

In practice, one would substitute an estimate of θ and the residual at time N into the above forecast formula.

4. Consider the ARIMA model on p. 92 where $p = 1$, $P = 0$, $q = 0$, and $Q = 1$. Show that

$$\hat{x}(N, 2) = x_{N-10} + \hat{\alpha}^2 (x_N - x_{N-12}) + \hat{\theta}\hat{\alpha}\hat{a}_{N-11}$$
$$+ \hat{\theta}\hat{a}_{N-10}.$$

5. For the Box–Jenkins seasonal ARIMA model with $p = 0$, $P = 1$, $q = 1$, $Q = 0$, and $w_t = \nabla_{12} x_t$, find forecasts at time N for up to 12 steps ahead in terms of observations and estimated residuals up to time N.

CHAPTER 6

Stationary Processes in the Frequency Domain

6.1 Introduction

In Chapter 3 we described several types of stationary stochastic process, placing emphasis on the autocovariance (or autocorrelation) function which is the natural tool for considering the evolution of a process through time. In this chapter we introduce a complementary function called the *spectral density function*, which is the natural tool for considering the frequency properties of a time series. Inference regarding the spectral density function is called an analysis in the *frequency domain*.

Some statisticians initially have difficulty in understanding the frequency approach, but the advantages of frequency methods are widely appreciated in such fields as electrical engineering, geophysics and meteorology. These advantages will become apparent in the next few chapters.

We shall confine ourselves to real-valued processes. Many authors consider the more general problem of complex-valued processes, and this results in some gain of mathematical conciseness. But, in my view, the reader is more likely to understand an approach restricted to real-valued processes. The vast majority of practical problems are covered by this approach.

110

6.2 The spectral distribution function

In order to introduce the idea of a spectral density function, we must first consider a function called the *spectral distribution function*. The approach adopted is heuristic and not mathematically rigorous, but will, hopefully, give the reader a better understanding of the subject than the more theoretical approach adopted for example by Grenander and Rosenblatt (1957).

Suppose we suspect that a time series contains a periodic component at a known frequency. Then a natural model is

$$X_t = R \cos(\omega t + \theta) + Z_t \tag{6.1}$$

where ω is called the *frequency* of the periodic variation, R is called the *amplitude* of the variation, θ is called the *phase*, and Z_t denotes some stationary random series. An example of the periodic component is shown in Fig. 6.1. Note that the angle $(\omega t + \theta)$ is measured in radians. Since

$$\omega = \text{number of radians per unit time}$$

it is sometimes called the *angular* frequency, but in keeping with most authors we will simply call ω the frequency. However some authors, notably Jenkins and Watts (1968), refer to frequency as

$$f = \omega/(2\pi)$$
$$= \text{number of cycles per unit time}$$

and this form of frequency is much easier to interpret from a physical point of view. We will usually use the angular frequency, ω, in mathematical formulae for conciseness, but will often use the frequency $f = \omega/(2\pi)$ for the interpretation of data. The period of a sinusoidal cycle, called the *wavelength*, is clearly $2\pi/\omega$ or $1/f$.

Model 6.1 is a very simple model but in practice the variation in a time series may be caused by variation at several different frequencies. For example, sales figures may contain weekly, monthly, yearly and other cyclical variation.

111

Figure 6.1 A graph of $R \cos(\omega t + \theta)$ with $R = 2$, $\omega = \pi/3$, and $\theta = \pi/6$.

In other words the data show variation at high, medium and low frequencies. It is natural therefore to generalize (6.1) to:

$$X_t = \sum_{j=1}^{k} R_j \cos(\omega_j t + \theta_j) + Z_t \tag{6.2}$$

where R_j is the amplitude at frequency ω_j.

The reader will notice that models (6.1) and (6.2) are *not* stationary if R, θ, $\{R_j\}$, and $\{\theta_j\}$ are fixed constants because $E(X_t)$ will change with time. In order to apply the theory of stationary processes to models like (6.1) and (6.2), it is customary to assume that R and $\{R_j\}$ are (uncorrelated) random variables with mean zero, or that θ and $\{\theta_j\}$ are random variables with a uniform distribution on $(0, 2\pi)$, which are fixed for a single realisation of the process (see Section 3.5 and exercise 14 of Chapter 3). This is something of a 'mathematical trick', but it does enable us to treat time series containing one or more deterministic sinusoidal components as stationary series.

Since $\cos(\omega t + \theta) = \cos \omega t \cos \theta - \sin \omega t \sin \theta$, model (6.2) is often expressed as a sum of sine and cosine terms in

112

the form

$$X_t = \sum_{j=1}^{k} (a_j \cos \omega_j t + b_j \sin \omega_j t) + Z_t \qquad (6.3)$$

where $a_j = R_j \cos \theta_j$

and $b_j = -R_j \sin \theta_j$.

But we may now ask why there should only be a finite number of frequencies involved in model (6.2) or (6.3). In fact, letting $k \to \infty$, the work of Wiener and others has shown that any discrete stationary process measured at unit intervals may be represented in the form

$$X_t = \int_0^\pi \cos \omega t \, du(\omega) + \int_0^\pi \sin \omega t \, dv(\omega) \qquad (6.4)$$

where $u(\omega)$, $v(\omega)$ are uncorrelated continuous processes with orthogonal increments (see Section 3.4.8) which are defined for all ω in the range $(0, \pi)$. Equation (6.4) is called the *spectral representation* of the process. The derivation of the spectral representation will not be considered here (see for example Cox and Miller, 1968, Chapter 8).

The reader may wonder why the upper limits of the integrals in (6.4) are π rather than ∞. For a continuous process, the upper limits would indeed be ∞, but for a discrete process measured at unit intervals of time there is no loss of generality in restricting ω to the range $(0, \pi)$ since

$$\cos(\omega t + k\pi t) = \begin{cases} \cos \omega t & k, t \text{ integers with } k \text{ even} \\ \cos(\pi - \omega)t & k, t \text{ integers with } k \text{ odd} \end{cases}$$

and so variation at frequencies higher than π cannot be distinguished from variation at the corresponding frequency in $(0,\pi)$. The frequency $\omega = \pi$, is called the *Nyquist frequency*. We will say more about this in Section 7.2.1. For a discrete process measured at equal intervals of time of length Δt, the Nyquist frequency is $\pi/\Delta t$. In the next two

113

sections we consider discrete processes measured at unit intervals of time, but the arguments carry over to discrete processes measured at intervals Δt if we replace π by $\pi/\Delta t$.

The main point of introducing the spectral representation (6.4) is to show that every frequency in the range $(0, \pi)$ may contribute to the variation of the process. However the processes $u(\omega)$ and $v(\omega)$ in (6.4) are of little direct practical interest. Instead we shall introduce a function, $F(\omega)$, called the (power) spectral distribution function, which is related to $u(\omega)$ and $v(\omega)$, and which arises from a theorem (e.g. Bartlett, 1966, Section 6.1), called the Wiener–Khintchine theorem, named after N. Wiener and A. Y. Khintchine. As applied to real-valued processes, this theorem says that for any stationary stochastic process with autocovariance function $\gamma(k)$, there exists a monotonically increasing function, $F(\omega)$, such that

$$\gamma(k) = \int_0^\pi \cos \omega k \, dF(\omega). \tag{6.5}$$

Equation (6.5) is called the spectral representation of the autocovariance function. It can be shown that the function, $F(\omega)$, has a direct physical interpretation, in that

$F(\omega)$ = contribution to the variance of the series which is accounted for by frequencies in the range $(0, \omega)$

It is most important to understand this physical interpretation of $F(\omega)$. There is no variation at negative frequencies so that

$$F(\omega) = 0 \quad \text{for} \quad \omega < 0$$

For a discrete process measured at unit intervals of time, the highest possible frequency is π and so all the variation is accounted for by frequencies less than π. Thus

$$F(\pi) = \text{Variance}(X_t) = \sigma_X^2.$$

This last result also comes directly from (6.5) with $k = 0$ when

$$\gamma(0) = \sigma_X^2 = \int_0^\pi dF(\omega).$$

In between $\omega = 0$ and $\omega = \pi$, $F(\omega)$ is monotonically increasing. In fact the function behaves rather like the cumulative distribution function of a probability distribution except that the upper limit is σ_X^2 rather than one.

If the process contains a deterministic sinusoidal component at frequency ω_0, say $R \cos(\omega_0 t + \theta)$ where R is a constant and θ is uniformly distributed on $(0, 2\pi)$, then there will be a step increase in $F(\omega)$ at ω_0 equal to $E[R^2 \cos^2(\omega_0 t + \theta)] = \frac{1}{2}R^2$.

As $F(\omega)$ is monotonic, it can be decomposed into two functions, $F_1(\omega)$ and $F_2(\omega)$ such that

$$F(\omega) = F_1(\omega) + F_2(\omega) \tag{6.6}$$

where $F_1(\omega)$ is a non-decreasing continuous function and $F_2(\omega)$ is a non-decreasing step function. This decomposition usually corresponds to the Wold decomposition with $F_1(\omega)$ relating to the purely indeterministic component of the process, while $F_2(\omega)$ relates to the deterministic component.

We shall be mainly concerned with purely indeterministic processes, where $F_2(\omega) \equiv 0$, so that $F(\omega)$ is a continuous function on $(0, \pi)$.

The adjective 'power', which is sometimes prefixed to 'spectral distribution function' derives from the engineer's use of the word in connection with the passage of an electric current through a resistance. For a sinusoidal input, the power is directly proportional to the squared amplitude of the oscillation. For a more general input, the power spectral distribution function describes how the power is distributed with respect to frequency. In the case of a time series, the variance may be regarded as the total power.

Note that some authors use a normalized form of $F(\omega)$

given by

$$F^*(\omega) = F(\omega)/\sigma_X^2 \tag{6.7}$$
$$= \textit{proportion} \text{ of variance accounted for by}$$
$$\text{frequencies in the range } (0, \omega).$$

Since $F^*(\pi) = 1$, and $F^*(\omega)$ is monotonically increasing, $F^*(\omega)$ has similar properties to a cumulative distribution function.

6.3 The spectral density function

For a purely indeterministic discrete stationary process, the spectral distribution function is a continuous (monotone bounded) function in $(0, \pi)$, and may therefore be differentiated† with respect to ω in $(0, \pi)$. We will denote the derivative by $f(\omega)$ where

$$f(\omega) = \frac{dF(\omega)}{d\omega} \tag{6.8}$$
$$= \text{(power) spectral density function.}$$

The term 'spectral density function' is often shortened to *spectrum*. The adjective 'power' is sometimes omitted.

When $f(\omega)$ exists, equation (6.5) can be expressed in the form

$$\gamma(k) = \int_0^\pi \cos \omega k \, f(\omega) d\omega \tag{6.9}$$

Putting $k = 0$, we then have

$$\gamma(0) = \sigma_X^2 = \int_0^\pi f(\omega) d\omega = F(\pi). \tag{6.10}$$

The physical meaning of the spectrum is that $f(\omega) \, d\omega$ represents the contribution to variance of components with

---

†(Strictly speaking, $F(\omega)$ may not be differentiable on a set of measure zero, but this is of no practical importance.)

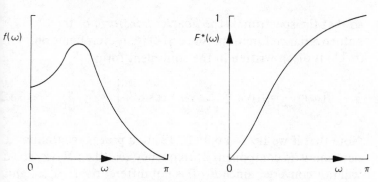

Figure 6.2   An example of a spectrum together with the corresponding normalized spectral distribution function.

frequencies in the range (ω, ω + dω). When the spectrum is drawn, equation (6.10) indicates that the total area underneath the curve is equal to the variance of the process. A peak in the spectrum indicates an important contribution to variance at frequencies in the appropriate region. An example of a spectrum is shown in Fig. 6.2 together with the corresponding normalized spectral distribution function.

It is important to realize that the autocovariance function (acv.f) and the power spectral density function are equivalent ways of describing a stationary stochastic process. From a practical point of view, they are complementary to each other. Both functions contain the same information but express it in different ways. In some situations a time-domain approach based on the acv.f. is more useful while in other situations a frequency-domain approach is preferable. The relationship between the two functions is somewhat similar to that between the probability density function and the characteristic function of a continuous probability distribution.

Equation (6.9) expresses $\gamma(k)$ in terms of $f(\omega)$ as a cosine transform. The inverse relationship (see Appendix I) is given by

$$f(\omega) = \frac{1}{\pi} \sum_{k=-\infty}^{\infty} \gamma(k) e^{-i\omega k} \qquad (6.11)$$

117

so that the spectrum is the *Fourier transform* of the autocovariance function. Since $\gamma(k)$ is an even function, (6.11) is often written in the equivalent form

$$f(\omega) = \frac{1}{\pi} \left[\gamma(0) + 2 \sum_{k=1}^{\infty} \gamma(k) \cos \omega k \right].$$ (6.12)

Note that if we try to apply (6.12) to a process containing a deterministic component at frequency ω_0, then $\Sigma \gamma(k) \cos \omega$ will not converge, since $F(\omega)$ is not differentiable at ω_0 and so $f(\omega_0)$ is not defined.

The reader should be made aware at this point that several other definitions of the spectrum are given in the literature, most of which differ from (6.12) by a constant multiple and by the range of definition of $f(\omega)$. The most popular approach is to define the spectrum in the range $(-\pi, \pi)$ by

$$f(\omega) = \frac{1}{2\pi} \sum_{k=-\infty}^{\infty} \gamma(k) e^{-i\omega k}$$ (6.13)

whose inverse relationship (see Appendix I) is

$$\gamma(k) = \int_{-\pi}^{\pi} e^{i\omega k} f(\omega) \, d\omega$$ (6.14)

Jenkins and Watts (1968) use these equations except that they take $f = \omega/2\pi$ as the variable (see equations A3 and A4 in Appendix I). Equations (6.13) and (6.14), which form a Fourier transform pair, are the more usual form of the Wiener–Khintchine relations. The formulation is slightly more general in that it can be applied to complex valued time series. But for real time series, we find that $f(\omega)$ is an even function and then one need only consider $f(\omega)$ for $\omega > 0$. In my experience, the introduction of negative frequencies, while having certain mathematical advantages, serves only to confuse the student. As we are concerned only with real-valued processes, we prefer (6.11) defined on $(0, \pi)$.

It is sometimes useful to use a normalized form of the

spectral density function given by

$$f^*(\omega) = f(\omega)/\sigma_X^2$$
$$= \frac{dF^*(\omega)}{d\omega} \tag{6.15}$$

This is the derivative of the normalized spectral distribution function (see equation 6.7). Then we find that $f^*(\omega)$ is the Fourier transform of the *autocorrelation* function, namely

$$f^*(\omega) = \frac{1}{\pi} \left[1 + 2 \sum_{k=1}^{\infty} \rho(k) \cos \omega k \right], \tag{6.16}$$

and that $f^*(\omega) \, d\omega$ is the *proportion* of variance between $(\omega, \omega + d\omega)$. Kendall and Stuart (1966) and Kendall (1976) define the spectral density function in the range $(0, \pi)$ in terms of the autocorrelation function but omit the constant $1/\pi$ from equation (6.16). This makes it more difficult to give the function a physical interpretation. [Note that the inverse relationship (6.56) given by Kendall (1976) is incorrect.]

6.4 The spectrum of a continuous process

For a continuous purely indeterministic stationary process, $X(t)$, the autocovariance function $\gamma(\tau)$ is defined for all τ and the (power) spectral density function, $f(\omega)$, is defined for all positive ω. The relationship between these functions is very similar to that in the discrete case except that there is no upper bound to the frequency. We have

$$f(\omega) = \frac{1}{\pi} \int_{-\infty}^{\infty} \gamma(\tau) e^{-i\omega\tau} \, d\tau$$
$$= \frac{2}{\pi} \int_{0}^{\infty} \gamma(\tau) \cos \omega\tau \, d\tau \tag{6.17}$$

for $0 < \omega < \infty$, with the inverse relationship

119

$$\gamma(\tau) = \int_0^\infty f(\omega) \cos \omega\tau \, d\omega. \tag{6.18}$$

The integral in (6.17) converges for all ω provided that $X(t)$ is a purely indeterministic process (except in a few pathological cases of no practical importance).

6.5 Examples

In this section we derive the spectral density functions of some simple but important stationary processes.

(a) Purely random process   A purely random process in discrete time, $\{Z_t\}$, is defined in Section 3.4.1. If $\mathrm{Var}(Z_t) = \sigma_Z^2$, then the acv.f. is given by

$$\gamma(k) = \begin{cases} \sigma_Z^2 & k = 0 \\ 0 & \text{otherwise} \end{cases}$$

so that the power spectral density function is given by

$$f(\omega) = \sigma_Z^2/\pi \tag{6.19}$$

using (6.12). In other words the spectrum is constant in the range $(0, \pi)$.

We have already pointed out that a continuous white noise process is physically unrealizable. A process is regarded as a practical approximation to continuous white noise if its spectrum is substantially constant over the frequency band of interest, even if it then approaches zero at high frequency.

(b) First-order moving average process   The first-order MA process (see Section 3.4.3)

$$X_t = Z_t + \beta Z_{t-1}$$

has ac.f.

$$\rho(k) = \begin{cases} 1 & k = 0 \\ \beta/(1 + \beta^2) & k = \pm 1 \\ 0 & \text{otherwise.} \end{cases}$$

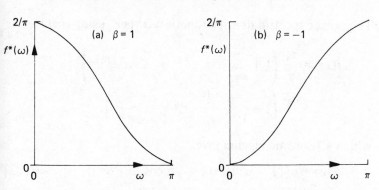

*Figure 6.3   Two examples of spectra of first-order moving average
processes with (a) $\beta = 1$; (b) $\beta = -1$.*

So, using (6.16), the normalized spectral density function is
given by

$$f^*(\omega) = \frac{1}{\pi} [1 + 2\beta \cos \omega / (1 + \beta^2)] \qquad (6.20)$$

for $0 < \omega < \pi$. The power spectral density function is then

$$f(\omega) = \sigma_X^2 f^*(\omega)$$

where $\sigma_X^2 = (1 + \beta^2)\sigma_Z^2$.

The shape of the spectrum depends on the value of β. When
$\beta > 0$, the power is concentrated at low frequencies giving
what is called a low-frequency spectrum, while if $\beta < 0$,
power is concentrated at high frequencies giving a high
frequency spectrum. Examples are shown in Fig. 6.3.

(c) First-order autoregressive process   The first-order AR
process (see Section 3.4.4)

$$X_t = \alpha X_{t-1} + Z_t \qquad (6.21)$$

has acv.f.

$$\gamma(k) = \sigma_X^2 \, \alpha^{|k|} \qquad (k = 0, \pm 1, \pm 2, \dots).$$

121

The power spectral density function is then, using (6.11),

$$f(\omega) = \frac{\sigma_X^2}{\pi} \left(1 + \sum_{k=1}^{\infty} \alpha^k e^{-ik\omega} + \sum_{k=1}^{\infty} \alpha^k e^{ik\omega} \right)$$

$$= \frac{\sigma_X^2}{\pi} \left(1 + \frac{\alpha e^{-i\omega}}{1 - \alpha e^{-i\omega}} + \frac{\alpha e^{i\omega}}{1 - \alpha e^{i\omega}} \right)$$

which after some algebra gives

$$f(\omega) = \sigma_X^2 (1 - \alpha^2)/[\pi(1 - 2\alpha \cos \omega + \alpha^2)] \qquad (6.22)$$

$$= \sigma_Z^2/[\pi(1 - 2\alpha \cos \omega + \alpha^2)] \qquad (6.23)$$

since $\sigma_Z^2 = \sigma_X^2 (1 - \alpha^2)$.

The shape of the spectrum depends on the value of α. When $\alpha > 0$, power is concentrated at low frequencies while if $\alpha < 0$ power is concentrated at high frequencies. Examples are shown in Fig. 6.4.

It is hoped that the reader finds the shapes of the spectra in Fig. 6.4 intuitively reasonable. For example if α is negative then it is clear from (6.21) that values of X_t will tend to oscillate: a positive value of X_t will tend to be followed by a negative value and vice versa. These rapid oscillations correspond to high frequency variation.

Figure 6.4   Two examples of spectra of first-order autoregressive processes with (a) $\alpha = 0.7$; (b) $\alpha = -0.7$.

122

(d) Higher-order autoregressive processes   It can be shown (e.g. Jenkins and Watts, 1968) that the spectrum of a second-order AR process with parameters α_1, α_2 is given by

$$f(\omega) = \sigma_Z^2 / \pi [1 + \alpha_1^2 + \alpha_2^2 - 2\alpha_1(1 - \alpha_2) \cos \omega$$
$$- 2\alpha_2 \cos 2\omega]$$

for $0 < \omega < \pi$. The shape of the spectrum depends on the values of α_1 and α_2. It is possible to get a high frequency spectrum, a low frequency spectrum, a spectrum with a peak between 0 and π, or a spectrum with a minimum between 0 and π.

For higher-order AR processes, one can get spectra with several peaks or troughs.

(e) A deterministic sinusoidal perturbation   Suppose that

$$X_t = \cos(\omega_0 t + \theta) \qquad (6.24)$$

where ω_0 is a constant in $(0, \pi)$ and θ is a random variable which is uniformly distributed on $(0, 2\pi)$. As explained in Section 3.5, θ is fixed for a single realization of the process and (6.24) defines a purely deterministic process.

The acv.f. of the process is given by

$$\gamma(k) = \frac{1}{2} \cos \omega_0 k$$

which we note does *not* tend to zero as k increases. This is a feature of most deterministic processes.

From (6.24), it is obvious that all the 'power' of the process is concentrated at the frequency ω_0. Now $\mathrm{Var}(X_t) = E(X_t^2) = \frac{1}{2}$, so that the power spectral distribution function is given by

$$F(\omega) = \begin{cases} 0 & \omega < \omega_0 \\ \frac{1}{2} & \omega \geqslant \omega_0 \end{cases}$$

Since this is a step function, it has no derivative at ω_0 and so the spectrum is not defined at ω_0. If we nevertheless try to

123

use equation (6.12) to obtain the spectrum as the Fourier transform of the acv.f. then we find that

$$f(\omega) = 0 \qquad \omega \neq \omega_0$$

but that $\Sigma\gamma(k) \cos \omega k$ does not converge at $\omega = \omega_0$. Jenkins and Watts (1968, p. 224) suggest writing the spectrum in terms of the Dirac delta function (see Appendix II), but there are mathematical difficulties involved in this approach.

(f) A mixture   Our final example contains a mixture of deterministic and stochastic components, namely

$$X_t = \cos(\omega_0 t + \theta) + Z_t$$

where ω_0, θ, are as defined in example (e) above, and $\{Z_t\}$ is a purely random process, mean zero and variance σ_Z^2. Then we find that the acv.f. is given by

$$\gamma(k) = \begin{cases} \frac{1}{2} + \sigma_Z^2 & k = 0 \\ \frac{1}{2} \cos \omega_0 k & k = \pm 1, \pm 2, \ldots \end{cases}.$$

Again note that $\gamma(k)$ does *not* tend to zero as k increases because X_t contains a deterministic component.

We can obtain the power spectral distribution function by using (6.6), as the deterministic component, $\cos(\omega_0 t + \theta)$, has distribution function

$$F_1(\omega) = \begin{cases} 0 & \omega < \omega_0 \\ \frac{1}{2} & \omega \geqslant \omega_0 \end{cases}$$

while the stochastic component, Z_t, has a distribution function

$$F_2(\omega) = \sigma_Z^2 \omega/\pi \qquad (0 < \omega < \pi)$$

on integrating (6.19). Thus the overall spectral distribution function is given by

$$F(\omega) = \begin{cases} \sigma_Z^2 \omega/\pi & 0 < \omega < \omega_0 \\ \frac{1}{2} + \sigma_Z^2 \omega/\pi & \omega_0 \leqslant \omega < \pi \end{cases}.$$

PROCESSES IN THE FREQUENCY DOMAIN

As in example (e), the power spectrum is not defined at $\omega = \omega_0$.

Exercises

In the following questions, $\{Z_t\}$ denotes a purely random process, mean zero and variance σ_Z^2.

1. Find (a) the power spectral density function, (b) the normalized spectral density function of the first-order AR process:

$$X_t = \lambda X_{t-1} + Z_t.$$

2. Find the power spectral density functions of the following MA processes:

   (a) $X_t = Z_t + Z_{t-1} + Z_{t-2}$
   (b) $X_t = Z_t + 0.5Z_{t-1} - 0.3Z_{t-2}$

3. Show that the second-order MA process

$$X_t = \mu + Z_t + 0.8Z_{t-1} + 0.5Z_{t-2}$$

   is second-order stationary, where μ denotes a constant. Find the acv.f. and ac.f. of $\{X_t\}$ and show that its normalized spectral density function is given by

$$f^*(\omega) = (1 + 1.27 \cos \omega + 0.53 \cos 2\omega)/\pi.$$
$$(0 < \omega < \pi)$$

4. A stationary time series $(X_t; t = \ldots, -1, 0, +1, \ldots)$ has normalized spectral density function

$$f^*(\omega) = 2(\pi - \omega)/\pi^2. \qquad (0 < \omega < \pi)$$

   Show that its ac.f. is given by

$$\rho(k) = \begin{cases} 1 & k = 0 \\ (2/\pi k)^2 & k \text{ is odd} \\ 0 & k \text{ is even } (\neq 0) \end{cases}$$

5. A two-state Markov process may be set up as follows. Alpha particles from a radioactive source are used to trigger a flip-flop device which takes the states $+1$ and -1 alternately. The times t_i at which changes occur constitute a Poisson process, with mean event rate λ. Let $X(t)$ denote the state variable at time t. If the process is started at $t = 0$ with $P(X(0) = 1) = P(X(0) = -1) = \frac{1}{2}$, show that the process is second-order stationary, with autocorrelation function

$$\rho(u) = e^{-2\lambda |u|} \qquad (-\infty < u < \infty)$$

and spectral density function

$$f(\omega) = 4\lambda/[\pi(4\lambda^2 + \omega^2)] \qquad (0 < \omega < \infty).$$

6. Show that if $\{X_t\}$ and $\{Y_t\}$ are independent, stationary processes with power spectral density functions $f_x(\omega)$ and $f_y(\omega)$, then $\{Z_t\} = \{X_t + Y_t\}$ is also stationary with power spectral density function $f_z(\omega) = f_x(\omega) + f_y(\omega)$. If

$$Z_t = X_t + Y_t$$

where

$$X_t = \alpha X_{t-1} + W_t \qquad (-1 < \alpha < +1)$$

and $\{Y_t\}, \{W_t\}$ are independent purely random processes with zero mean and common variance σ^2, show that the power spectral density function of $\{Z_t\}$ is given by

$$f_z(\omega) = \sigma^2(2 - 2\alpha \cos \omega + \alpha^2)/\pi(1 - 2\alpha \cos \omega + \alpha^2)$$
$$(0 < \omega < \pi).$$

7. Show that the power spectral density function of the ARMA $(1,1)$ process

$$X_t = \alpha X_{t-1} + Z_t + \beta Z_{t-1}$$

is given by

$$f(\omega) = \sigma_z^2(1 + 2\beta \cos \omega + \beta^2)/\pi(1 - 2\alpha \cos \omega$$
$$+ \alpha^2) \quad (0 < \omega < \pi)$$

CHAPTER 7

Spectral Analysis

Spectral analysis, sometimes called 'spectrum analysis', is the name given to methods of *estimating* the spectral density function, or spectrum, of a given time series.

In the last century, research workers, such as A. Schuster, were essentially concerned with looking for 'hidden periodicities' in the data, but spectral analysis as we know it today is mainly concerned with estimating the spectrum over the whole range of frequencies. M. S. Bartlett and J. W. Tukey have been prominent in the development of modern spectral analysis which has a history of about thirty years, and the techniques are now widely used by many scientists, particularly in electrical engineering, physics, meteorology and marine science.

We are mainly concerned with purely indeterministic processes, which have a continuous spectrum, but the techniques can also be used for deterministic processes to pick out periodic components in the presence of noise.

7.1 Fourier analysis

Spectral analysis is essentially a modification of *Fourier* analysis so as to make it suitable for stochastic rather than deterministic functions of time. Fourier analysis (e.g. Hsu, 1967) is basically concerned with approximating a function by a sum of sine and cosine terms, called the Fourier series representation. Suppose that a function $f(t)$ is defined on $(-\pi, \pi]$ and satisfies the Dirichlet conditions [i.e. that it is absolutely integrable over this range, has a finite number of discontinuities, and a finite number of maxima and minima]. Then $f(t)$ may be approximated by the Fourier series

$$\frac{a_0}{2} + \sum_{r=1}^{k} (a_r \cos rt + b_r \sin rt),$$

where $a_0 = \dfrac{1}{\pi} \displaystyle\int_{-\pi}^{\pi} f(t)\mathrm{d}t$

$$a_r = \frac{1}{\pi} \int_{-\pi}^{\pi} f(t) \cos rt \, \mathrm{d}t \qquad (r = 1, 2, \dots)$$

$$b_r = \frac{1}{\pi} \int_{-\pi}^{\pi} f(t) \sin rt \, \mathrm{d}t. \qquad (r = 1, 2, \dots).$$

It can be shown that this Fourier series converges to $f(t)$ as $k \to \infty$ except at points of discontinuity where it converges to $\frac{1}{2}[f(t+0) + f(t-0)]$.

In order to apply Fourier analysis to discrete time series, we need to consider the Fourier series representation of $f(t)$ when $f(t)$ is defined only on the integers $1, 2, \dots, N$. Rather than write down the formula, we shall see that the required Fourier series emerges naturally by considering a simple sinusoidal model.

7.2 A simple sinusoidal model

Suppose we suspect that a given time series, with observations made at unit time intervals, contains a deterministic sinusoidal component at frequency ω together with a random error term. So we will consider the model

$$X_t = \mu + \alpha \cos \omega t + \beta \sin \omega t + Z_t \tag{7.1}$$

where Z_t denotes a purely random process, and μ, α, β are parameters to be estimated from the data.

The observations will be denoted by $(x_1, x_2, \dots, x_N)$. The algebra in the next few sections is somewhat simplified if we confine ourselves to the case where N is *even*. Some other authors (e.g. Jenkins and Watts, 1968) also impose this condition. There is no real difficulty in extending the results to the case where N is odd (e.g. Anderson, 1971), and indeed many of the later estimation formulae apply for both odd and even N, but some results require one to consider odd N and even N separately. For example it is a little disconcerting to find that the Fourier series representation of a discrete time series with N odd (e.g. Pocock, 1974) does not contain a component at the Nyquist frequency. Thus, if N happens to be odd and a spectral analysis is required, our preference is to remove the first observation so as to make N even. If N is reasonably large, little information is lost and there is some gain in simplicity.

The model (7.1) can be represented in matrix notation by

$$E(\mathbf{X}) = A\boldsymbol{\theta}$$

where

$$\mathbf{X}^T = (X_1, \ldots, X_N)$$
$$\boldsymbol{\theta}^T = (\mu, \alpha, \beta)$$

and
$$A = \begin{pmatrix} 1 & \cos \omega & \sin \omega \\ 1 & \cos 2\omega & \sin 2\omega \\ \cdot & \cdots\cdots & \cdots\cdots \\ 1 & \cos N\omega & \sin N\omega \end{pmatrix}$$

It is 'well-known' that the least-squares estimate of $\boldsymbol{\theta}$ which minimizes $\sum_{t=1}^{N} (x_t - \mu - \alpha \cos \omega t - \beta \sin \omega t)^2$ is given by (e.g. Jenkins and Watts, 1968, Appendix A4.1)

$$\hat{\boldsymbol{\theta}} = (A^T A)^{-1} A^T \mathbf{x}$$

where

$$\mathbf{x}^T = (x_1, \ldots, x_N).$$

Now the highest frequency we can fit to the data is the Nyquist frequency, given by $\omega = \pi$ (see Section 7.2.1), while the lowest frequency we can reasonably fit completes 1 cycle in the whole length of the time series (see Section 7.2.1). By equating the cycle length $2\pi/\omega$ to N, we find that this lowest frequency is given by $2\pi/N$. The least-squares estimates are particularly simple if ω is restricted to one of the values

$$\omega_p = 2\pi p/N \qquad (p = 1, \ldots , N/2)$$

as $(A^T A)$ then turns out to be a diagonal matrix in view of the following 'well-known' trigonometric results (all summations are for $t = 1$ to N):

$$\Sigma \cos \omega_p t = \Sigma \sin \omega_p t = 0 \tag{7.2}$$

$$\Sigma \cos \omega_p t \cos \omega_q t = \begin{cases} 0 & p \neq q \\ N & p = q = N/2 \\ N/2 & p = q \neq N/2 \end{cases} \tag{7.3}$$

$$\Sigma \sin \omega_p t \sin \omega_q t = \begin{cases} 0 & p \neq q \\ 0 & p = q = N/2 \\ N/2 & p = q \neq N/2 \end{cases} \tag{7.4}$$

$$\Sigma \cos \omega_p t \sin \omega_q t = 0 \qquad \text{for all } p, q. \tag{7.5}$$

With $(A^T A)$ diagonal, we can easily find $\hat{\boldsymbol{\theta}}$, and for ω_p ($p \neq N/2$) we find (exercise 2)

$$\left. \begin{array}{l} \hat{\mu} = \Sigma x_t/N = \bar{x} \\ \hat{\alpha} = 2\Sigma x_t \cos(\omega_p t)/N \\ \hat{\beta} = 2\Sigma x_t \sin(\omega_p t)/N \end{array} \right\} \tag{7.6}$$

while if $p = N/2$ we ignore the term in $\beta \sin \omega t$ which is zero for all t and find

$$\left. \begin{array}{l} \hat{\mu} = \bar{x} \\ \hat{\alpha} = \Sigma x_t(-1)^t/N \end{array} \right\} \tag{7.7}$$

The model (7.1) is essentially the one used in the last century to search for hidden periodicities, but this model has now gone out of fashion. However it can still be useful if we have reason to suspect that a time series does contain a deterministic periodic component at a known frequency and we wish to isolate this component (e.g. Granger and Hatanaka, 1964, p. 6). Pocock (1974) gives an example where this type of procedure is used to investigate the seasonal variation in sickness absence.

Readers who are familiar with the analysis of variance technique will see that the total corrected sum of squared deviations, namely

$$\sum_{t=1}^{N} (x_t - \bar{x})^2,$$

can be partitioned into two components which are the residual sum of squares and the sum of squares 'explained' by the periodic component at frequency ω_p. This latter component is given by

$$\sum_{t=1}^{N} (\hat{\alpha} \cos \omega_p t + \hat{\beta} \sin \omega_p t)^2$$

which after some algebra (exercise 2) can be shown to be

$$\begin{matrix} (\hat{\alpha}^2 + \hat{\beta}^2)N/2 & \quad p \neq N/2 \\ \hat{\alpha}^2 N & \quad p = N/2 \end{matrix} \tag{7.8}$$

using (7.2)–(7.5).

7.2.1 The Nyquist frequency   In Chapter 6 we pointed out that for a discrete process measured at unit intervals there is no loss of generality in restricting the spectral distribution function to the range $(0, \pi)$. We now demonstrate that the upper bound, π, called the Nyquist frequency, is indeed the highest frequency about which we can get meaningful information from a set of data.

First we will give a more general form for the Nyquist frequency. If observations are taken at equal intervals of length Δt, then the Nyquist (angular) frequency is $\omega_N = \pi/\Delta t$. The equivalent frequency expressed in cycles per unit time is $f_N = \omega_N/2\pi = 1/2\Delta t$.

Consider the following example. Suppose that temperature readings are taken every day in a certain town at noon. It is clear that these observations will tell us nothing about temperature variation *within* a day. In particular they will not tell us if nights are hotter or cooler than days. With only one observation per day, the Nyquist frequency is $\omega_N = \pi$ radians per day or $f_N = 1$ cycle per two days. This is lower than the frequencies which account for variation within a day. For example variation with a wavelength of one day has (angular) frequency $\omega = 2\pi$ radians per day or $f =$ one cycle per day. In order to get information about variation within a day we must take two or more observations per day.

A similar example is provided by yearly sales figures which will obviously give no information about any seasonal effects, whereas monthly or quarterly observations *will* give information about seasonality.

Finally a comment about the lowest frequency we can fit to a set of data. If we had just six months of temperature readings from winter to summer it would not be clear if there was an upward trend in the observations or if winters are colder than summers. However with one year's data it *would* become clear that winters are colder than summers. Thus if we are interested in variations within a year at the comparatively low frequency of 1 cycle per year, then we must have at least one year's data. Thus the lower the frequency we are interested in, the longer the time period over which we need to take measurements, whereas the higher the frequency we are interested in, the more frequently must we take observations.

7.3 Periodogram analysis

Early attempts at discovering hidden periodicities in a given time series basically consisted of repeating the analysis of Section 7.2 at all the frequencies $2\pi/N$, $4\pi/N \ldots$, π. In view of (7.3)–(7.5) the different terms are orthogonal and we end up with the finite Fourier series representation of the $\{x_t\}$, namely

$$x_t = a_0 + \sum_{p=1}^{N/2-1} [a_p \cos(2\pi pt/N) + b_p \sin(2\pi pt/N)]$$
$$+ a_{N/2} \cos \pi t, \qquad (t = 1, 2, \ldots, N) \tag{7.9}$$

where the coefficients $\{a_p, b_p\}$ are of the same form as equations (7.6) and (7.7), namely

$$\left. \begin{array}{l} a_0 = \bar{x} \\ a_{N/2} = \Sigma(-1)^t x_t/N \\ \left. \begin{array}{l} a_p = 2[\Sigma x_t \cos(2\pi pt/N)]/N \\ b_p = 2[\Sigma x_t \sin(2\pi pt/N)]/N \end{array} \right\} \ p = 1, \ldots, N/2-1 \end{array} \right\} \tag{7.10}$$

An analysis along these lines is sometimes called a Fourier analysis or a *harmonic* analysis. The Fourier series representation (7.9) has N parameters to describe N observations and so can be made to fit the data exactly (just as a polynomial of degree $(N-1)$ involving N parameters can be found which goes exactly through N observations in polynomial regression). This explains why there is no error term in (7.9) in contrast to (7.1). Also note that there is no term in $\sin \pi t$ in (7.9) as $\sin \pi t$ is zero for all integer t.

It is worth stressing that the Fourier series coefficients (7.10) at a given frequency ω are exactly the same as the least-squares estimates for model (7.1).

The overall effect of the Fourier analysis of the data is to partition the variability of the series into components at frequencies $2\pi/N$, $4\pi/N, \ldots$, π. The component at frequency

$\omega_p = 2\pi p/N$ is often called the pth harmonic. For $p \neq N/2$, it is often useful to write the pth harmonic in the equivalent form

$$a_p \cos \omega_p t + b_p \sin \omega_p t = R_p \cos (\omega_p t + \phi_p) \qquad (7.11)$$

where

R_p = *amplitude* of pth harmonic

$$= \sqrt{(a_p^2 + b_p^2)} \qquad (7.12)$$

and ϕ_p = *phase* of pth harmonic

$$= \tan^{-1} (-b_p/a_p). \qquad (7.13)$$

We have already noted in Section 7.2 that, for $p \neq N/2$, the contribution of the pth harmonic to the total sum of squares is given by $N(a_p^2 + b_p^2)/2$. Using (7.12), this is equal to $NR_p^2/2$. Extending this result using (7.2)–(7.5) and (7.9), we have, after some algebra, that (exercise 3)

$$\sum_{t=1}^{N} (x_t - \bar{x})^2 = N \sum_{p=1}^{N/2-1} R_p^2/2 + Na_{N/2}^2. \qquad (7.14)$$

Dividing through by N we have

$$\Sigma(x_t - \bar{x})^2/N = \sum_{p=1}^{N/2-1} R_p^2/2 + a_{N/2}^2, \qquad (7.15)$$

which is often known as Parseval's theorem. The left-hand side of (7.15) is effectively the variance of the observations although the divisor is N rather than the more usual $(N-1)$. Thus $R_p^2/2$ is the contribution of the pth harmonic to the variance, and (7.15) shows how the total variance is partitioned.

If we plot $R_p^2/2$ against $\omega_p = 2\pi p/N$ we obtain a line spectrum. A different type of line spectrum occurs in the physical sciences when light from molecules in a gas discharge tube is viewed through a spectroscope. The light has energy at discrete frequencies and this energy can be seen as bright

lines. But most time series have continuous spectra and then
it is inappropriate to plot a line spectrum. If we regard $R_p^2/2$
as the contribution to variance in the range $\omega_p \pm \pi/N$, we can
plot a histogram whose height in the range $\omega_p \pm \pi/N$ is such
that

$R_p^2/2$ = area of histogram rectangle

= height of histogram × $2\pi/N$.

Thus the height of the histogram is given by

$$I(\omega_p) = NR_p^2/4\pi. \tag{7.16}$$

As usual, (7.16) does not apply for $p = N/2$; we may regard
$a_{N/2}^2$ as the contribution to variance in the range
$[\pi(N-1)/N, \pi]$ so that

$$I(\pi) = N a_{N/2}^2/\pi.$$

The plot of $I(\omega)$ against ω is usually called the
periodogram even though $I(\omega)$ is a function of frequency
rather than period. Other authors define the periodogram in a
slightly different way, as some other multiple of R_p^2. For
example Granger and Hatanaka (1964, equation 1.2.3) define
a quantity which is π × expression (7.16). [Note that these
authors use ω in equation 1.2.3 to denote period, while in
the rest of their book ω denotes frequency.] Hannan (1970,
equation 3.8) defines the periodogram in terms of complex
numbers as

$$\frac{1}{2\pi N} \left| \sum_{t=1}^{N} x_t e^{it\omega} \right|^2$$

which is ½ × expression (7.16). Anderson (1971,
Section 4.3.2) describes the graph of R_p^2 against the *period*
N/p as the periodogram, and suggests the term *spectrogram*
to describe the graph of R_p^2 against frequency. The advantage
of definition (7.16) is that the total area under the
periodogram is equal to the variance of the time series.
Expression (7.16) may readily be calculated directly from the

135

data by

$$I(\omega_p) = [(\Sigma x_t \cos 2\pi pt/N)^2 + (\Sigma x_t \sin 2\pi pt/N)^2]/N\pi.$$

$$(7.17)$$

Equation (7.17) also applies for $p = N/2$. Jenkins and Watts (1968) define a similar expression in terms of the variable $f = \omega/2\pi$, but call it the 'sample spectrum'.

The periodogram appears to be a natural way of estimating the power spectral density function, but we shall see that for a process with a *continuous* spectrum it provides a poor estimate and needs to be modified.

7.3.1 The relationship between the periodogram and the autocovariance function
The periodogram ordinate, $I(\omega)$, and the autocovariance coefficient, c_k, are both quadratic forms of the data $\{x_t\}$. It is of interest to see how they are related. We will show that the periodogram is the finite Fourier transform of $\{c_k\}$.

Using (7.2) we may rewrite (7.17) for $p \neq N/2$ as

$$I(\omega_p) = [(\Sigma(x_t - \bar{x}) \cos \omega_p t)^2 + (\Sigma(x_t - \bar{x}) \sin \omega_p t)^2]/$$

$$= \sum_{s,t=1}^{N} (x_t - \bar{x})(x_s - \bar{x})(\cos \omega_p t \cos \omega_p s$$

$$+ \sin \omega_p t \sin \omega_p s) / N\pi$$

But $\sum_{t=1}^{N-k} (x_t - \bar{x})(x_{t+k} - \bar{x})/N = c_k \qquad$ [see (4.1)]

and $\cos \omega_p t \cos \omega_p(t + k) + \sin \omega_p t \sin \omega_p(t + k) = \cos \omega_p k$, so that

$$I(\omega_p) = (c_0 + 2 \sum_{k=1}^{N-1} c_k \cos \omega_p k)/\pi \qquad (7.18)$$

$$= \sum_{k=-(N-1)}^{N-1} c_k e^{-i\omega_p k} /\pi. \qquad (7.19)$$

Equation (7.18) is perhaps the more useful relationship

connecting the periodogram and autocovariance function. The equivalent form (7.19) is the more familiar form of a finite Fourier transform assuming that $c_k = 0$ for $|k| \geqslant N$.

7.3.2 Properties of the periodogram

When the periodogram is expressed in the form (7.18), it appears to be the 'obvious' estimate of the power spectrum

$$f(\omega) = (\gamma_0 + 2 \sum_{k=1}^{\infty} \gamma_k \cos \omega k)/\pi$$

simply replacing γ_k by its estimate c_k for values of k up to $(N-1)$, and putting subsequent estimates of γ_k equal to zero. But although we find

$$\underset{N \to \infty}{E} (I(\omega)) \to f(\omega) \qquad\qquad (7.20)$$

so that the periodogram is asymptotically unbiased, we will see that the variance of $I(\omega)$ does *not* decrease as N increases. Thus $I(\omega)$ is *not* a *consistent* estimator for $f(\omega)$. An example of a periodogram is given in Fig. 7.5(c) on p. 161 and it can be seen that the graph fluctuates wildly. The lack of consistency is perhaps not too surprising when one realizes that the Fourier series representation (7.9) requires one to evaluate N parameters from N observations however long the series. Thus in Section 7.4 we will consider alternative ways of estimating a power spectrum which are essentially ways of *smoothing* the periodogram.

We complete this section by proving that $I(\omega)$ is not a consistent estimator for $f(\omega)$ in the case where $(x_1, \ldots, x_N)$ are taken from a discrete purely random process, where the observations are independent $N(\mu, \sigma^2)$ variates. It can also be shown that this result holds for other stationary processes with continuous spectra (e.g. Bartlett, 1966, p. 308) but this will not be demonstrated here.

From (7.10) we see that a_p and b_p are linear combinations of normally distributed random variables and so will themselves be normally distributed. Using (7.2)–(7.4), it can be shown (exercise 4) that a_p and b_p each have mean zero and variance $2\sigma^2/N$ for $p \neq N/2$. Furthermore we have

$$\text{Covariance}(a_p, b_p) = 4 \text{ Cov} \{(\Sigma x_t \cos \omega_p t), (\Sigma x_t \sin \omega_p t)\}/N^2$$

$$= 4 \sigma^2 (\Sigma \cos \omega_p t \sin \omega_p t)/N^2$$

since the $\{x_t\}$ are independent. Thus, using (7.5), we see that a_p and b_p are uncorrelated. Since (a_p, b_p) are bivariate normal, zero correlation implies that a_p and b_p are independent. Now a result from distribution theory says that if Y_1, Y_2 are independent $N(0, 1)$ variables, then $(Y_1^2 + Y_2^2)$ has a χ^2 distribution with 2 degrees of freedom, which is written χ_2^2. Thus

$$\frac{N(a_p^2 + b_p^2)}{2\sigma^2} = \frac{I(\omega_p) \times 2\pi}{\sigma^2}$$

is χ_2^2. Now the variance of a χ^2 distribution with ν degrees of freedom is 2ν, so that

$$\text{Var}[I(\omega_p) \times 2\pi/\sigma^2] = 4$$

and

$$\text{Var}[I(\omega_p)] = \sigma^4/\pi^2$$

As this variance is a constant, it does *not* tend to zero as $N \to \infty$, and hence $I(\omega_p)$ is not a consistent estimator for $f(\omega_p)$. Furthermore it can be shown that neighbouring periodogram ordinates are asymptotically independent, which further explains the very irregular form of an observed periodogram. Thus the periodogram needs to be modified in order to obtain a good estimate of a continuous spectrum.

7.4 Spectral analysis: some consistent estimation procedures

This section describes several alternative procedures for carrying out a spectral analysis. The different methods will be compared in Section 7.6. Each method provides a *consistent* estimate of the (power) spectral density function, in contrast to the periodogram. But although the periodogram is itself an inconsistent estimate, we shall see that the procedures described in this section are essentially based on the periodogram by using some sort of smoothing procedure.

Throughout the section we will assume that any obvious trend or seasonal variation has been removed from the data. If this is not done, the results of the spectral analysis are likely to be dominated by these effects, making any other effects difficult or impossible to see. Trend will produce a peak at zero frequency, while seasonal variation produces peaks at the seasonal frequency and at integer multiples of the seasonal frequency. These integer multiples of the fundamental frequency are called *harmonics* (see Section 7.8). Godfrey (1974) has pointed out that, for non-stationary series, it is often true that the final results depend more on the method chosen to remove trend than on any other factor.

The methods described in this Chapter are essentially non-parametric in that no model is assumed *a priori*. Recently a parametric approach called autoregressive spectrum estimation has been proposed and this will be introduced in Chapter 10.

7.4.1 Transforming the truncated autocovariance function

A popular type of estimation procedure consists of taking a Fourier transform of the truncated sample autocovariance function using a weighting procedure. From equation (7.18), we have that the periodogram is the discrete Fourier transform of the complete sample autocovariance function. But it is clear that the precision of the c_k decreases as k increases, so that it would seem intuitively reasonable to give less weight to the values of c_k as k increases. An estimator which has this property is:

$$f(\omega) = \frac{1}{\pi} \{\lambda_0 c_0 + 2 \sum_{k=1}^{M} \lambda_k c_k \cos \omega k\} \tag{7.21}$$

where $\{\lambda_k\}$ are a set of weights called the *lag window*, and

139

$M(< N)$ is called the *truncation point.* Comparing (7.21) with (7.18) we see that values of c_k for $M < k < N$ are no longer used, while values of c_k for $k \leqslant M$ are weighted by a factor λ_k.

In order to use the above estimator, the reader must choose a suitable lag window and a suitable truncation point. The two lag windows which are in common use today are:

(a) Tukey Window

$$\lambda_k = \tfrac{1}{2}(1 + \cos \frac{\pi k}{M}) \quad k = 0, 1, \ldots M.$$

This window is also called the Tukey–Hanning or Blackman–Tukey window.

(b) Parzen Window

$$\lambda_k = \begin{cases} 1 - 6\left(\dfrac{k}{M}\right)^2 + 6\left(\dfrac{k}{M}\right)^3 & 0 \leqslant k \leqslant M/2 \\ 2(1 - k/M)^3 & M/2 \leqslant k \leqslant M. \end{cases}$$

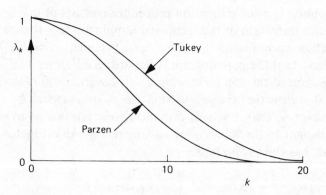

Figure 7.1   The Tukey and Parzen lag windows with M = 20.

These two windows are illustrated in Figure 7.1 with $M = 20$.

The Tukey and Parzen windows will give very much the same estimated spectrum for a given time series, although the Parzen window has a slight advantage in that it cannot give negative estimates. Many other lag windows have been suggested (see Hannan, 1970, Section 5.4) and 'window carpentry' was a popular research topic in the 1950s. Ways of comparing different windows will be discussed in Section 7.6. The well-known Bartlett window, with $\lambda_k = 1 - k/M$ for $k = 0, 1, \ldots, M$, is no longer used as its properties are inferior to the Tukey and Parzen windows. Neave (1972b) has suggested an alternative window which has superior properties but which is more complicated to use.

The choice of the truncation point, M, is rather difficult and little clear-cut advice is available in the literature. It has to be chosen subjectively so as to balance 'resolution' against 'variance'. The smaller the value of M, the smaller will be the variance of $\hat{f}(\omega)$ but the larger will be the bias (Neave, 1971). If M is too small, important features of $f(\omega)$ may be smoothed out, while if M is too large the behaviour of $\hat{f}(\omega)$ becomes more like that of the periodogram with erratic variation. Thus a compromise value must be chosen. A useful rough guide is to choose M to be about $2\sqrt{N}$, so that if for example N is 200, then M will be round about the value 28. This choice of M ensures the asymptotic situation that as $N \to \infty$, so also does $M \to \infty$ but in such a way that $M/N \to 0$. A somewhat larger value of M is required for the Parzen window than for the Tukey window. Jenkins and Watts (1968) suggest trying 3 different values of M. A low value will give an idea where the large peaks in $f(\omega)$ are, but the curve is likely to be too smooth. A high value is likely to produce a curve showing a large number of peaks, some of which may be spurious. A compromise can then be achieved with the third value of M. As Hannan (1970, p. 311) says, 'experience is the real teacher and that cannot be got from a book'.

In principle (7.21) may be evaluated at any value of ω in $(0, \pi)$, but it is usually evaluated at equal intervals at $\omega = \pi j/Q$ for $j = 0, 1, \ldots, Q$, where Q is chosen sufficiently large to show up all features of $\hat{f}(\omega)$. Often Q is chosen to be equal to M. The graph of $\hat{f}(\omega)$ against ω can then be plotted and examined. An example is given on p. 161 for the data plotted in Fig. 1.2 using the Tukey window with $M = 24$.

7.4.2 Hanning   This procedure, named after Julius Von Hann, is equivalent to the use of the Tukey window as described in Section 7.4.1, but adopts a different computational procedure. The estimated spectrum is calculated in two stages. Firstly a truncated unweighted cosine transform of the data is taken to give

$$\hat{f}_1(\omega) = \frac{1}{\pi} \left(c_0 + 2 \sum_{k=1}^{M} c_k \cos \omega k \right). \tag{7.22}$$

This is the same as (7.21) except that the lag window is taken to be unity (i.e. $\lambda_k = 1$). The estimates given by (7.22) are calculated at $\omega = \pi j/M$ for $j = 0, 1, \ldots, M$. These estimates are then smoothed using the weights ($\frac{1}{4}, \frac{1}{2}, \frac{1}{4}$) to give the Hanning estimates.

$$\hat{f}(\omega) = \tfrac{1}{4}\hat{f}_1(\omega - \pi/M) + \tfrac{1}{2}\hat{f}_1(\omega) + \tfrac{1}{4}\hat{f}_1(\omega + \pi/M) \tag{7.23}$$

at $\omega = \pi j/M$ for $j = 1, 2, \ldots, (M - 1)$. At zero frequency, and at the Nyquist frequency, π, we take

$$\hat{f}(0) = \tfrac{1}{2}[\hat{f}_1(0) + \hat{f}_1(\pi/M)]$$

$$\hat{f}(\pi) = \tfrac{1}{2}[\hat{f}_1(\pi) + \hat{f}_1(\pi(M - 1)/M)]$$

It is easily demonstrated that this procedure is equivalent to the use of the Tukey window. Substituting (7.22) into (7.23) we find

$$\hat{f}(\omega) = \frac{1}{\pi} \left\{ c_0 + 2 \sum_{k=1}^{M} c_k [\tfrac{1}{4}\cos(\omega - \pi/M)k + \tfrac{1}{2}\cos \omega k \right.$$
$$\left. + \tfrac{1}{4}\cos(\omega + \pi/M)k] \right\}$$

But $\cos(\omega - \pi/M)k + \cos(\omega + \pi/M)k = 2 \cos \omega k \cos(\pi k/M)$ and comparing with (7.21) we see that the lag window is indeed the Tukey window.

There is relatively little difference in the computational efficiency of Hanning and the straightforward use of the Tukey window. Both methods yield the same estimates and so it matters little which of the two procedures is used in practice.

7.4.3 Hamming   This technique is very similar to Hanning and has a very similar title which sometimes leads to confusion. In fact Hamming is named after a quite different person, namely R. W. Hamming. The technique is nearly identical to Hanning except that the weights (¼, ½, ¼) in (7.23) are changed to (0.23, 0.54, 0.23). At the frequencies $\omega = 0$ and $\omega = \pi$, the weights are 0.54 and 0.46. The procedure gives similar estimates to those produced by Hanning.

7.4.4 Smoothing the periodogram   The methods of Sections (7.4.1)–(7.4.3) are based on transforming the sample autocovariance function. An alternative type of approach is to smooth the periodogram by simply grouping the periodogram ordinates in sets of size m and finding their average value.[†] This approach is based on a suggestion by P. J. Daniell in 1946. Then we find

$$\hat{f}(\omega) = \frac{1}{m} \sum_j I(\omega_j) \qquad (7.24)$$

where $\omega_j = 2\pi j/N$ and j varies over m consecutive integers so that the ω_j are symmetric about ω. In order to estimate $f(\omega)$ at $\omega = 0$ and $\omega = \pi$, (7.24) has to be modified in an obvious way, treating the periodogram as being symmetric about 0 and π. Then, taking m to be odd with $m^* = (m-1)/2$, we have

$$\hat{f}(0) = 2 \sum_{j=1}^{m^*} I(2\pi j/N)/m$$

[†] It is also possible to smooth the periodogram ordinates by a variety of non-uniform averaging procedures.

THE ANALYSIS OF TIME SERIES

assuming $I(0) = 0$. We also have

$$\hat{f}(\pi) = \left[I(\pi) + 2 \sum_{j=1}^{m} {}^* I(\pi - 2\pi j/N) \right] /m.$$

Now we know that the periodogram is asymptotically unbiased but inconsistent for the true spectrum. Since neighbouring periodogram ordinates are asymptotically uncorrelated, it is clear that the variance of (7.24) will be of order $1/m$. It is also clear that the estimator (7.24) may be biased since

$$E(\hat{f}(\omega)) \simeq \frac{1}{m} \sum_{j} f(\omega_j)$$

which is equal to $f(\omega)$ only if the spectrum is linear over the interval. However the bias will be unimportant provided that $f(\omega)$ is a reasonably smooth function at ω and m is not too large compared with N.

Thus the choice of m is rather like the choice of the truncation point, M, in Section 7.4.1 in that it has to be chosen so as to balance resolution against variance, although the effects are in opposite directions. The larger the value of m the smaller will be the variance of the resulting estimate but the larger will be the bias, and if m is too large then interesting features of $f(\omega)$, such as peaks, may be smoothed out. As N increases, so we can allow m to increase.

There seems to be relatively little advice in the literature on the choice of m. It seems advisable to try several values, in the region of $N/40$. A high value should give some idea where the large peaks in $f(\omega)$ are, but the curve is likely to be too smooth. A low value is likely to produce a curve showing a large number of peaks, some of which may be spurious. A compromise can then be made.

Although the procedure described in this section is computationally quite different from that of Section 7.4.1, there are in fact close links between the two procedures. In

Section 7.3.1 we derived the relationship between the periodogram and the sample autocovariance function, and if we substitute equation (7.18) into (7.24) we can express the estimate, $\hat{f}(\omega)$, in terms of the sample autocovariance function in a similar form to equation (7.21). We find, after some algebra (exercise 5), that the truncation point is $(N - 1)$ and the lag window is given by

$$\lambda_k = \begin{cases} 1 & k = 0 \\[2mm] \dfrac{\sin(m\pi k/N)}{m \sin(\pi k/N)} & k = 1, 2, \ldots, N-1 \end{cases}$$

Historically, the method was not much used until recent years because it apparently requires much more computational effort than the procedure of Section 7.4.1. Calculating the periodogram using equation (7.17) at ω_p for $p = 1, 2, \ldots, N/2$, would require about N^2 arithmetic operations (each one a multiplication and an addition), whereas using equation (7.21) less than MN operations are required to calculate the $\{c_k\}$ so that the total number of operations is only of order $M(N + M)$ if $Q = M$. Two factors have led to the increasing use of the smoothed periodogram. Firstly the advent of high speed computers means that it is no longer necessary to restrict oneself to the method requiring the fewest calculations (Jones, 1965). For example when $N = 1000$, it takes less than a minute to compute the periodogram on an IBM 7094 using equation (7.17). The second factor has been the rediscovery of a technique called the Fast Fourier Transform which can speed up the computation of the periodogram quite considerably. This technique will now be described.

7.4.5 The fast Fourier transform   This technique substantially reduces the time required to perform a Fourier analysis on a computer, and is also more accurate. The title is usually abbreviated to FFT and we will use this abbreviation. (But note that Hannan (1970) uses this abbreviation to denote *finite* Fourier transform.)

A history of the FFT is described by Cooley *et al.* (1967), the ideas going back to the early 1900's. But it was the work of J. W. Cooley, J. W. Tukey and G. Sande about 1965 which first stimulated the application of the technique to time-series analysis. Much of the subsequent work has been published in IEEE Transactions on Audio and Electroacoustics (now renamed IEEE Trans. on Acoustics, Speech and Signal Processing) which devoted two issues in June, 1967, and June, 1969, entirely to this topic. We will only give a broad outline of the technique here. For further details, see for example Bingham *et al.* (1967), Otnes and Enochson (1972, Chapter 4), Bendat and Piersol (1971), and Bloomfield (1976).

The basic idea of the FFT can be illustrated in the case when N can be factorized in the form $N = rs$. If we assume that N is even, then at least one of the factors, say r, will be even. Using complex numbers for mathematical simplicity, the Fourier coefficients from equation (7.10) are given by

$$a_p + ib_p = 2[\Sigma x_t \exp(2\pi ipt/N)]/N \qquad (7.25)$$

for $p = 0, 1, 2, \ldots, N/2 - 1$. For mathematical convenience we denote the observations by $x_0, x_1, \ldots, x_{N-1}$ so that the summation in (7.25) is from $t = 0$ to $(N - 1)$. Now we can write t in the form

$$t = rt_1 + t_0$$

where $t_1 = 0, 1, \ldots, (s - 1)$, and $t_0 = 0, 1, \ldots, (r - 1)$, as t goes from 0 to $(N - 1)$, in view of the fact that $N = rs$. Similarly we can decompose p in the form

$$p = sp_1 + p_0$$

where $p_1 = 0, 1, \ldots, (r/2 - 1)$, and $p_0 = 0, 1, \ldots; (s - 1)$, as p goes from 0 to $(N/2 - 1)$. Then the summation in (7.25) may be written

$$\sum_{t_0=0}^{r-1} \exp(2\pi i p t_0/N) \sum_{t_1=0}^{s-1} x_t \exp(2\pi i prt_1/N).$$

But $\exp(2\pi i prt_1/N) = \exp(2\pi i (sp_1 + p_0)rt_1/N) = \exp(2\pi i p_0 rt_1/N)$

since $\exp(2\pi i sp_1 rt_1/N) = \exp(2\pi i p_1 t_1) = 1$ for all p_1, t_1.

Thus $\sum_{t_1=0}^{s-1} x_t \exp(2\pi i prt_1/N)$

does not depend on p_1 and is therefore a function of t_0 and p_0 only, say $A(p_0, t_0)$. Then (7.25) may be written

$$a_p + ib_p = 2 \left[\sum_{t_0=0}^{r-1} A(p_0, t_0) \exp(2\pi i p t_0/N) \right] \Big/ N.$$

Now there are rs functions of type $A(p_0, t_0)$ to be calculated, each requiring s complex multiplications and additions. Then the $(a_p + ib_p)$ may be calculated with $(r^2 s/2)$ complex multiplications and additions giving a grand total of $rs(s + r/2) = N(s + r/2)$ calculations instead of the $N^2/2$ calculations required to use (7.25) directly.

Much bigger reductions in computing can be made by an extension of the above procedure when N is highly composite (i.e. has many small factors). In particular if N is of the form 2^n, then we find that the number of operations is of order Nn(or $N \log_2 N$) instead of N^2. Substantial gains can also be made when N has several factors (e.g. $N = 2^p 3^q 5^r \dots$) and Singleton (1969) gives a general computer program.

In practice it is unlikely that N will be of a simple form such as 2^n, although it may be possible to make N highly composite by omitting a few observations. More generally we can add zeros to the (mean-corrected) sample record so as to *increase* the value of N until it is a suitable integer. Then a procedure called *tapering* or data windowing (e.g. Tukey, 1967; Godfrey, 1974; Bloomfield, 1976) is sometimes recommended to avoid a discontinuity at the end of the data, though its use is now controversial. Suppose for example

147

that we happen to have 382 observations. This value of N is *not* highly composite and we might proceed as follows:

(a) Remove any linear trend from the data, and keep the residuals (which should have mean zero) for subsequent analysis. If there is no trend, simply subtract the overall mean from each observation.

(b) Apply a linear taper to about 5% of the data at each end. In this example, if we denote the detrended mean-corrected data by $x_0, x_1, \ldots, x_{381}$, then the tapered series is given by

$$x_t{}^* = \begin{cases} (t+1)x_t/20 & t = 0, 1, \ldots, 18 \\ (382-t)x_t/20 & t = 363, \ldots, 381 \\ x_t & t = 19, 20, \ldots, 362 \end{cases}$$

(c) Add $(512-382) = 130$ zeros at one end of the tapered series, so that $N = 512 = 2^9$.

(d) Carry out an FFT on the data, calculate the Fourier coefficients $a_p + ib_p$, and average the values of $(a_p^2 + b_p^2)$ in groups of about 10.

In fact with N as low as 382, the computational advantage of the FFT is limited and we could equally well calculate the periodogram directly which avoids the need for tapering and adding zeros. The FFT really comes into its own for $N > 1000$.

Jenkins and Watts (1968) give two reasons why they think the case for using the FFT in spectral analysis is not strong. Firstly they say that fast computers are more than adequate for carrying out a spectral analysis by traditional methods. This is certainly true for $N < 1000$, but with several thousand observations even fast computers will take several minutes without the FFT. Secondly Jenkins and Watts say that the autocorrelation function is an invaluable intermediate stage in spectral analysis. This is also true, but does not mean that the FFT is useless because it is generally quicker to calculate the sample autocovariance function by performing *two*

FFT's. As described by Bingham *et al.* (1967), we can compute the Fourier coefficients (a_p, b_p) with an FFT of the mean-corrected data at $\omega_p = 2\pi p/N$ for $p = 0, 1, \ldots, N - 1$ and not for $p = 0, 1, \ldots, N/2$ as we usually do. The extra coefficients are redundant for real-valued processes since $a_{N-k} = a_k$ and $b_{N-k} = -b_k$. We then compute $R_p^2 = a_p^2 + b_p^2$ and Fast Fourier retransform the sequence (R_p^2) to get the mean lagged products. With $N = 2^{12} = 4096$, R. Fenton found that it took over three times as long to calculate one sixth of the $\{c_k\}$ directly than to calculate *all* the $\{c_k\}$ using two FFT's. In using the FFT for this purpose, one has to be careful to add enough zeros to the data (*without* tapering) to make sure that one gets non-circular sums of lagged products, as defined by equation (4.1) and used throughout this book. Circular coefficients result if zeros are not added where, for example, the circular autocovariance coefficient at lag 1 is

$$c_1^* = \left[\sum_{t=1}^{N} (x_t - \bar{x})(x_{t+1} - \bar{x}) \right]/N$$

where x_{N+1} is taken equal to x_1. If $x_1 = \bar{x}$, the circular and non-circular coefficients at lag 1 are the same. To calculate all the autocovariance coefficients of a set of N observations one adds N zeros, to make $2N$ 'observations' in all.

7.5 Confidence intervals for the spectrum

The methods of Section 7.4 all produce *point* estimates of the spectral density function at different frequencies. In this section we show how to find confidence intervals for the spectrum at different frequencies.

In Section 7.3.2 we showed that a white noise process, with constant spectrum $f(\omega) = \sigma^2/\pi$, has a periodogram ordinate at frequency ω, $I(\omega)$, which is such that $2I(\omega)/f(\omega)$ is distributed as χ_2^2. A more general result is given by Jenkins and Watts (1968, Section 6.4.2) for the estimator of

Section 7.4.1, namely

$$\hat{f}(\omega) = \left[\sum_{k=-M}^{M} \lambda_k c_k \cos \omega k \right] / \pi \, ,$$

which is that asymptotically $\nu\hat{f}(\omega)/f(\omega)$ is approximately distributed as χ_ν^2, where

$$\nu = 2N \Big/ \sum_{k=-M}^{M} \lambda_k^2 \qquad (7.26)$$

is called the number of degrees of freedom of the lag window. Then

$$P(\chi_{\nu,1-\alpha/2}^2 < \nu\hat{f}(\omega)/f(\omega) < \chi_{\nu,\alpha/2}^2) = 1 - \alpha$$

so that the $100(1 - \alpha)\%$ confidence interval for $f(\omega)$ is given by

$$\frac{\nu\hat{f}(\omega)}{\chi_{\nu,\alpha/2}^2} \text{ to } \frac{\nu\hat{f}(\omega)}{\chi_{\nu,1-\alpha/2}^2} \, .$$

The degrees of freedom for the Tukey and Parzen windows turn out to be $2.67N/M$ and $3.71N/M$ respectively. When smoothing the periodogram in groups of size m, it is clear that the result will have $2m$ degrees of freedom and there is no need to apply equation (7.26). In fact equation (7.26) does not work for the periodogram when expressed in the form (7.18) as noted by Hannan (1970, p. 281).

The confidence intervals given in this section are asymptotic results. Neave (1972a) has shown that these results are also quite accurate for short series.

7.6 A comparison of different estimation procedures

Several factors need to be considered when comparing the different estimation procedures which were described in Section 7.4. These include such practical considerations as computing time and the availability of computer programs.

We begin by considering the theoretical properties of the different procedures. Other comparative discussions are given by Jones (1965), Jenkins and Watts (1968), Neave (1972b), and Bloomfield (1976).

It is useful to introduce a function called the *spectral window* or *kernel* which is the Fourier transform of the lag window. If we define the lag window, λ_k, to be zero for $> M$, and to be symmetric so that $\lambda_{-k} = \lambda_k$, then the spectral window is given by

$$K(\omega) = \frac{1}{2\pi} \sum_{k=-\infty}^{\infty} \lambda_k \, e^{-ik\omega} \qquad (7.27)$$

or $(-\pi < \omega < \pi)$. This has the inverse relation

$$\lambda_k = \int_{-\pi}^{\pi} K(\omega) \, e^{i\omega k} \, d\omega. \qquad (7.28)$$

All the estimation procedures for the spectrum can be put in the general form

$$\hat{f}(\omega_0) = \frac{1}{\pi} \sum_{k=-N+1}^{N-1} \lambda_k \, c_k \, e^{-i\omega_0 k}$$

$$= \frac{1}{\pi} \sum \left[\int_{-\pi}^{\pi} K(\omega) e^{i\omega k} \, d\omega \right] c_k \, e^{-i\omega_0 k}$$

$$= \frac{1}{\pi} \int_{-\pi}^{\pi} K(\omega) \, [\Sigma \, c_k \, e^{ik(\omega - \omega_0)}] \, d\omega$$

$$= \int_{-\pi}^{\pi} K(\omega) \, I(\omega_0 - \omega) d\omega \qquad (7.29)$$

using equation (7.19). Equation (7.29) shows that all the estimation procedures are essentially smoothing the periodogram using the weight function $K(\omega)$. The value of the lag window at lag zero is usually specified to be one, so that from (7.28) we have

$$\lambda_0 = 1 = \int_{-\pi}^{\pi} K(\omega)d\omega$$

which is a desirable property for a smoothing function.

Taking expectations in equation (7.29) we have asymptotically that

$$E[\hat{f}(\omega_0)] = \int_{-\pi}^{\pi} K(\omega) f(\omega_0 - \omega)d\omega. \qquad (7.30)$$

Thus the spectral window is a weight function expressing the contribution of the spectral density function at each frequency to the expectation of $\hat{f}(\omega_0)$. The name 'window' arises from the fact that $K(\omega)$ determines the part of the periodogram which is 'seen' by the estimator.

Examples of the spectral windows for the three commonest methods of spectral analysis are shown in Fig. 7.2, which is adapted from Jones (1965, Fig. 5). Taking $N = 1000$, the spectral window for the smoothed periodogram with $m = 20$ is shown as line A. The other two windows are the Parzen and Tukey windows, denoted by lines B and C. The values of the truncation point, M, were chosen to be 93 for the Parzen window and 67 for the Tukey window. These values of M were choosen so that all three windows gave estimators with equal variance. Formulae for variances will be given later in this section.

Inspecting Fig. 7.2, we see that the Parzen and Tukey windows look very similar, although the Parzen window has the advantage of being non-negative and of having smaller side-lobes. The shape of the periodogram window is quite different. It is approximately rectangular with a sharp cut-off and is close to the 'ideal' band-pass filter which would be exactly rectangular but which is unattainable in practice. The periodogram window also has the advantage of being non-negative. Overall we can say that the periodogram has a

Figure 7.2   The spectral windows for three common methods of spectral analysis; A: Smoothed periodogram (m = 20); B: Parzen (M = 93); C: Tukey (M = 67); all with N = 1000.

window with more desirable properties than those of Parzen and Tukey.

In comparing different windows, we also want to consider *bias, variance* and *bandwidth*. We will not derive formulae for the bias produced by the different procedures. It is clear from equation (7.30) that the wider the window, the larger will be the bias. In particular it is clear that all the smoothing procedures will tend to lower peaks and raise troughs.

As regards variance, we have from Section 7.5 that $\nu \hat{f}(\omega)/f(\omega)$ is approximately distributed as χ_ν^2, where $\nu = 2m$, $3.71\ N/M$, and $8N/3M$ for the smoothed periodogram, Parzen and Tukey windows, respectively. But

$$\text{Var}(\chi_\nu^2) = 2\nu$$

and

$$\text{Var}(\nu \hat{f}(\omega)/f(\omega)) = \nu^2\, \text{Var}(\hat{f}(\omega)/f(\omega))$$

153

and so Var($\hat{f}(\omega)/f(\omega)$) turns out to be $1/m$, $2M/3.71\,N$, and $3M/4N$ for the three windows. Equating these expressions gives the values of M chosen for Fig. 7.2.

Finally let us introduce the term bandwidth, which roughly speaking is the width of the spectral window. Various definitions are given in the literature, but we shall adopt that used by Jenkins and Watts (1968), namely the width of the 'ideal' rectangular window which would give an estimator with the same variance. The window of the smoothed periodogram is so close to being rectangular for m 'large' that it is clear from Fig. 7.2 that the bandwidth will be $2\pi m/N$. The bandwidths for the Parzen and Tukey windows turn out to be $2\pi \times 1.86/M$ and $8\pi/3M$ respectively. When plotting a graph of an estimated spectrum, it is a good idea to indicate the bandwidth which has been used.

The choice of bandwidth, or equivalently the choice of m or M, is an important step in spectral analysis. For the Parzen and Tukey windows, the bandwidth is inversely proportional to M (see Fig. 7.3). As M gets larger, the window gets narrower and the bias gets smaller but the variance of the resulting estimator gets larger. In fact the variance is inversely proportional to the bandwidth. For the smoothed periodogram, the bandwidth is directly proportional to m.

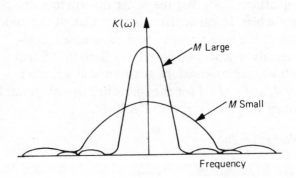

Figure 7.3   Spectral windows for different values of M.

For the unsmoothed periodogram, with $m = 1$, the window is very tall and narrow giving an estimator with large variance as we have already shown. All in all, the choice of bandwidth is rather like the choice of class interval when constructing a histogram.

We will now summarize the comparative merits of the three main estimation procedures. From the preceding discussion, it appears that the smoothed periodogram has superior theoretical properties in that its spectral window is approximately rectangular. From a computing point of view, it can be much slower for large N unless the FFT is used. If the FFT is used however, it can be much quicker and it is also possible to calculate the autocorrelation function using the FFT. For small N, computing time is relatively unimportant. Regarding computer programs, it is much easier to write a program for the Parzen or Tukey windows, but programs for the FFT are becoming more readily available and this trend will no doubt continue. In the future the use of the smoothed periodogram will probably become more general.

7.7 Analysing a continuous time series

We have so far been concerned with the spectral analysis of discrete time series. But time series are sometimes recorded as a continuous trace, as for example air temperature, the moisture content of tobacco emerging from a processing plant, and humidity. For series which contain components at very high frequencies, such as those arising in acoustics and speech processing, it may be possible to analyse them mechanically using tuned filters, but the more usual procedure is to *digitize* the series by reading off the values of the trace at discrete intervals. If values are taken at equal time intervals of length Δt, we have converted a continuous time series into a standard discrete time series and can use the methods already described.

In sampling a continuous time series, the main question is how to choose the sampling interval Δt. It is clear that sampling leads to some loss of information and that this loss gets worse as Δt increases. However it is expensive to make Δt very small and so a compromise value must be sought.

For the sampled series, the Nyquist frequency is $\pi/\Delta t$ radians per unit time, and we can get no information about variation at higher frequencies. Thus we clearly want to choose Δt so that variation in the continuous series is negligible at frequencies higher than $\pi/\Delta t$. In fact most measuring instruments are *band-limited* in that they do not respond to frequencies higher than a certain maximum frequency. If this maximum frequency is known or can be guessed, then the choice of Δt is straightforward.

If Δt is chosen to be too large, then a phenomenon called *aliasing* may occur. This can be illustrated by the following theorem.

Theorem 7.1 A continuous time series, with spectrum $f_c(\omega)$ for $0 < \omega < \infty$, is sampled at equal intervals of length Δt. The resulting discrete time series has spectrum $f_d(\omega)$ defined over $0 < \omega < \pi/\Delta t$. Then $f_d(\omega)$ and $f_c(\omega)$ are related by

$$f_d(\omega) = \sum_{s=0}^{\infty} f_c(\omega + 2\pi s/\Delta t) + \sum_{s=1}^{\infty} f_c(-\omega + 2\pi s/\Delta t) \quad (7.31)$$

Proof The proof will be given for the case $\Delta t = 1$. The extension to other values of Δt is straightforward.

The acv.f's of the continuous and sampled series will be denoted by $\gamma(\tau)$ and γ_k. It is clear that when τ takes an integer value, say k, then

$$\gamma(k) = \gamma_k. \quad (7.32)$$

Now from (6.9) we have

$$\gamma_k = \int_0^{\pi} f_d(\omega) \cos \omega k \, d\omega$$

156

while from (6.18) we have

$$\gamma(\tau) = \int_0^\infty f_c(\omega) \cos \omega\tau \, d\omega.$$

Thus, using (7.32), we have

$$\int_0^\pi f_d(\omega) \cos \omega k \, d\omega = \int_0^\infty f_c(\omega) \cos \omega k \, d\omega$$

for $k = 0, \pm1, \pm2, \ldots$. Now

$$\int_0^\infty f_c(\omega) \cos \omega k \, d\omega = \sum_{s=0}^\infty \int_{2\pi s}^{2\pi(s+1)} f_c(\omega) \cos \omega k \, d\omega$$

$$= \sum_{s=0}^\infty \int_0^{2\pi} f_c(\omega + 2\pi s) \cos \omega k \, d\omega.$$

$$= \sum_{s=0}^\infty \int_0^\pi \{f_c(\omega + 2\pi s) + f_c[2\pi(s+1) - \omega]\} \cos \omega k \, d\omega$$

$$= \int_0^\pi \left\{ \sum_{s=0}^\infty f_c(\omega + 2\pi s) + \sum_{s=1}^\infty f_c(2\pi s - \omega) \right\} \cos \omega k \, d\omega,$$

and the result follows.

The implications of this theorem are now considered. Firstly, we note that if the continuous series contains no variation at frequencies above the Nyquist frequency, so that $f_c(\omega) = 0$ for $\omega > \pi/\Delta t$, then $f_d(\omega) = f_c(\omega)$. In this case no information is lost by sampling. But more generally, the effect of sampling will be that variation at frequencies above the Nyquist frequency will be 'folded back' and produce an effect at a frequency lower than the Nyquist frequency in $f_d(\omega)$. If we denote the Nyquist frequency, $\pi/\Delta t$, by ω_N, then the frequencies $\omega, 2\omega_N - \omega, 2\omega_N + \omega, 4\omega_N - \omega, \ldots$, are called *aliases* of one another. Variation at all these frequencies in the continuous series will appear as variation at frequency ω in the sampled series.

From a practical point of view, aliasing will cause trouble

unless Δt is chosen so that $f_c(\omega) \simeq 0$ for $\omega > \pi/\Delta t$. If we have no advance knowledge about $f_c(\omega)$ then we can guesstimate a value for Δt. If the resulting estimate of $f_d(\omega)$ approaches zero near the Nyquist frequency $\pi/\Delta t$, then our choice of Δt was almost certainly sufficiently small. But if $f_d(\omega)$ does not approach zero near the Nyquist frequency, then it is probably wise to try a smaller value of Δt. Alternatively one can filter the continuous series to remove the high frequency components if one is interested in the low frequency components.

7.8 Discussion

Spectral analysis can be a useful exploratory diagnostic tool in the analysis of many types of time series. In this section we discuss how the estimated spectrum should be interpreted, when it is likely to be most useful and when it is likely to be least useful. We also discuss some of the practical problems arising in spectral analysis.

We begin this discussion with an example to give the reader some feel for the sorts of spectrum shape that may arise. Fig. 7.4 shows four sections of trace, labelled A, B, C and D, which were produced by four different processes.† The figure also shows the corresponding long-run spectra, labelled J, K, L and M, but these are given in *random* order. Note that the four traces use the same scale, the length produced in one second being shown on trace D. The four spectra are plotted using the same linear scales. The peak in spectrum L is at 15 cycles per second (or 15 Hz). The reader is invited to decide which series goes with which spectrum before reading on.

Trace A is much smoother than the other three traces. Its spectrum is therefore concentrated at low frequency and is actually spectrum M. The other three spectra are much harder to distinguish. Trace B is somewhat smoother than C or D and corresponds to spectrum K which 'cuts off' at a lower frequency than J or L. Trace C corresponds to

†Generated in a control engineering laboratory.

158

Figure 7.4   Four time series and their spectra. The spectra are given in random order.

spectrum J, while trace D contains a deterministic sinusoidal component at 15 cycles per second which contributes 20% of the total power. Thus D corresponds to spectrum L.

From a visual inspection of traces C and D, it is difficult or impossible to decide which goes with spectrum L. For this type of data, spectral analysis is invaluable in assessing the frequency properties. The reader may find it surprising that the deterministic component in trace D is so hard to see. In contrast the regular seasonal variation in air temperature at Recife given in Fig. 1.2 is quite obvious from a visual inspection of the time plot, but there the deterministic component accounts for 85% of the total variation. A spectral analysis of air temperature at Recife yields the spectrum shown in Fig. 7.5(a) with a large peak at a frequency of one cycle per year. But here the spectral analysis is not really necessary as the seasonal effect is obvious anyway. In fact if one has a series containing an obvious trend or seasonality, then such variation should be removed from the data *before* carrying out a spectral analysis, as any other effects will be relatively small and are unlikely to be visible in the spectrum of the raw data. Fig. 7.5(b) shows the spectrum of the Recife air temperature data when the seasonal variation has been removed. The variance is concentrated at low frequencies indicating either a trend, which is not apparent in Fig. 1.2, or short-term correlation as in a first-order AR process with a positive coefficient (c.f. Fig. 6.4a).

Removing trend and seasonality is the simplest form of a general procedure called *prewhitening*. It is easier to estimate the spectrum of a series which has a relatively flat spectrum. Prewhitening consists of making a linear transformation of the raw data so as to achieve a smoother spectrum, estimating the spectrum of the transformed data, and then using the transfer function of the linear transformation to estimate the spectrum of the raw data (see Chapter 9 and Anderson, 1971, p. 546). But this procedure requires some prior knowledge of

Figure 7.5   *Spectra for average monthly air temperature*
readings at Recife; (a) for the raw data; (b) for the
seasonally adjusted data using the Tukey window
with M = 24. *(c) The periodogram of the seasonally*
adjusted data is shown for comparison.

*Figure 7.6   The typical spectral
shape of an economic time-series.*

the spectral shape and is not often used except for removing
trend and seasonality.

Having estimated the spectrum of a given time series, how
do we interpret the results? There are various features to look
for. Firstly, are there any peaks in the spectrum? If so, why?
Secondly, is the spectrum large at low frequency, indicating
possible non-stationarity in the mean? Thirdly what is the
general shape of the spectrum? The typical shape of the power
spectrum of an economic variable is shown in Fig. 7.6, and
the implications of this shape are discussed by Granger
(1966). There are exceptions. Granger and Hughes (1971)
found a peak at a frequency of 1 cycle per 13 years when
they analysed Beveridge's yearly wheat price index series. But
this series is much longer than most economic series, and the
results have, in any case, been queried (Akaike, 1978).

The general shape of the spectrum may occasionally be
helpful in indicating an appropriate parametric model, but it
is not generally used in this way. The spectrum is not, for
example, used in the Box–Jenkins procedure for identifying
an appropriate ARIMA process (though Hannan, 1970,
Section 6.5 has suggested that it might be). Spectral analysis
is essentially a non-parametric procedure in which a finite set
of observations is used to estimate a function defined over
the range $(0, \pi)$. The function is not constrained to any

particular parametric class and so spectral analysis is a more
general procedure than inference based on a particular
parametric class of models, but is also likely to be less
accurate if a parametric model really is appropriate.

Spectral analysis is at its most useful for series of the type
shown in Fig. 7.4, with no obvious trend or 'seasonal'
variation. Such series arise mostly in the physical sciences. In
economics, spectral techniques have perhaps not proved as
useful as was first hoped, although there have been some
successes. Recently, attempts have been made to apply
spectral analysis to marketing data but it can be argued
(Chatfield, 1974) that marketing series are usually too short
and the seasonal variation too large for spectral analysis to
give useful results. In meteorology and oceanography,
spectral analysis can be very useful (e.g. Munk *et al.*, 1959;
Craddock, 1965; Roden, 1966; Snodgrass *et al.*, 1966;
Mooers and Smith, 1968) but even in these sciences, spectral
analysis often produces no worthwhile results. Chatfield and
Pepper (1971) analysed a number of monthly geophysical
series but found no tendency to oscillate at frequencies
other than the obvious annual effect.

We conclude this section by commenting on some practical
aspects of spectral analysis.

Most aspects, such as the choice of truncation point, have
already been discussed and will be further clarified in the
example given in Section 7.9. One problem, which has not
been discussed, is whether to plot the estimated spectrum or
its logarithm. An advantage of plotting the estimated
spectrum on a logarithmic scale is that its asymptotic
variance is then independent of the level of the spectrum so
that confidence intervals for the spectrum are of constant
width on a logarithmic scale. For spectra showing large
variations in power, a logarithmic scale also makes it possible
to show more detail over a wide range. (For example in
measuring sound, engineers use decibels which are measured
on a logarithmic scale.) Jenkins and Watts (1968. p. 266)

suggest that spectrum estimates should always be plotted on a logarithmic scale. But Anderson (1971, p. 547) points out that this exaggerates the visual effects of variations where the spectrum is small. Thus it is often easier to interpret a spectrum plotted on an arithmetic scale as the area under the graph corresponds to power and one can more readily assess the importance of different peaks. So, while it is often useful to plot $\hat{f}(\omega)$ on a logarithmic scale in the initial stages of a spectral analysis, when trying different truncation points and testing the significance of peaks, this writer generally prefers to plot the estimated spectrum on an arithmetic scale in order to interpret the final result. It is also generally easier to interpret a spectrum if the frequency scale is measured in cycles per unit time (f) rather than radians per unit time (ω). This has been done in Figs. 7.4 and 7.5. A linear transformation of frequency does not affect the *relative* heights of the spectrum at different frequencies which are of prime importance, though it does change the absolute heights by a constant multiple.

Another point worth mentioning is the possible presence in estimated spectra of *harmonics*. When a spectrum has a large peak at some frequency ω', then related peaks may occur at $2\omega, 3\omega, \ldots$. These multiples of the fundamental frequency are called harmonics and generally speaking simply indicate the non-sinusoidal character of the main cyclical component. For example Mackay (1973) studied the incidence of trips to supermarkets by consumers and found (not surprisingly!) a basic weekly pattern with harmonics at two and three cycles per week.

Finally, a question that is often asked is how large a value of N is required to get a reasonable estimate of the spectrum. It is often recommended that between 100 and 200 observations is the minimum. Granger and Hughes (1968) have tried smaller values of N and conclude that only very large peaks can then be found. If the data is prewhitened, so that the spectrum is fairly flat, then reasonable estimates may be obtained even with values of N less than 100.

7.9 An example

As an example, we analysed part of trace D of Fig. 7.4. Although a fairly long trace was available, we decided just to analyse a section lasting for one second to illustrate the problems of analysing a fairly short series. This set of data will also illustrate the problems of analysing a continuous trace as opposed to a discrete series.

The first problem was to digitize the data, and this required the choice of a suitable sampling interval. Inspection of the original trace showed that variation seemed to be 'fairly smooth' over a length of 1 mm corresponding to 1/100 second, but to ensure that there was no aliasing, we chose 1/200 second as the sampling interval, giving $N = 200$ observations.

For such a short series, there is little to be gained by using the Fast Fourier Transform. We decided to transform the truncated autocovariance function, using equation (7.21), with the Tukey window. Several truncation points were tried, and the results for $M = 20$, 40 and 80 are shown in Fig. 7.7. Above about 70 cycles per second, the estimates produced by the three values of M are very close to one another when the spectrum is plotted on an arithmetic scale and cannot be distinguished on the graph. Equation (7.21) was evaluated at 51 points, taking $Q = 50$, at $\omega = \pi j/50$ ($j = 0, 1, \ldots, 50$) where ω is measured in radians per unit time. Now in this example 'unit time' is 1/200 second and so the values of ω in radians per second are $\omega = 200\pi j/50$ for $j = 0, 1, \ldots, 50$. Thus the frequencies expressed in cycles per second, by $f = \omega/2\pi$, are $f = 2j$ for $j = 0, 1, \ldots, 50$. The Nyquist frequency is given by $f_N = 100$ cycles per second, which completes one cycle every two observations.

Figure 7.7 Estimated spectra for graph D of Fig. 7.4 using the Tukey window with (a) M = 80; (b) M = 40; (c) M = 20.

Looking at Fig. 7.7, the estimated spectrum is judged rather too smooth when $M = 20$, and much too erratic when $M = 80$. The value $M = 40$ looks about right although $M = 30$ might be even better. There is a clear peak at about 15 cycles per second (15 Hz) as there is in Fig. 7.4 (L), but there is also a smaller unexpected peak around 30 cycles per second, which looks like a harmonic of the variation at 15 cycles per second. If a longer series of observations were to be analysed, this peak might disappear indicating that the peak is spurious.

We also estimated the spectrum using a Parzen window with a truncation point of $M = 56$. This value was chosen so that the degrees of freedom of the window, namely 13.3, was almost the same as for the Tukey window with $M = 40$. I intended to plot the results on Fig. 7.7, but the graph was so close to that produced by the Tukey window that it was impossible to draw them both on the same graph. The biggest difference in estimates at the same frequency was 0.33 at 12 cycles per second but most of the estimates differed only in the second decimal place. It is clear that the Tukey and Parzen windows give very much the same estimates when equivalent values of M are used.

One feature to note in Fig. 7.7 is that the estimated spectrum approaches zero as the frequency approaches the Nyquist frequency. This suggests that there is no aliasing and that our choice of sampling interval was sufficiently small.

Also note that the bandwidths are indicated in Fig. 7.7. The bandwidth for the Tukey window is $8\pi/3M$ in radians per unit time. As 'unit time' is $1/200$ second, the bandwidth is $1600\pi/3M$ in radians per second or $800/3M$ in cycles per second.

Confidence intervals can be calculated as described in Section 7.5. For a sample of only 200 observations, they turn out to be disturbingly wide. For example when $M = 40$, the degrees of freedom are $2.67N/M = 13.3$. For convenience this is rounded off to the nearest integer, namely $\nu = 13$. The peak in the estimated spectrum is at 14 cycles per second where

$\hat{f}(\omega)$ = 7.5. Here the 95% confidence interval is (3.9 to 19.5). Clearly a longer series is desirable to make the confidence intervals acceptably narrow.

Exercises

1. Revision of Fourier series. Show that the Fourier series which represents the function

$$f(x) = x^2 \qquad \text{in } -\pi \leqslant x \leqslant \pi$$

is given by

$$f(x) = \frac{\pi^2}{3} - 4 \left[\frac{\cos x}{1} - \frac{\cos 2x}{2^2} + \frac{\cos 3x}{3^2} - \cdots \right]$$

2. Derive equations (7.6) and (7.8).

3. Derive Parseval's theorem given by equation (7.15).

4. If $X_1, \ldots, X_N$ are independent $N(\mu, \sigma^2)$ variates, show that

$$a_p = 2[\Sigma X_t \cos(2\pi pt/N)]/N$$

is $N(0, 2\sigma^2/N)$ for $p = 1, 2, \ldots, N/2 - 1$.

5. Derive the lag window for smoothing the periodogram in sets of size m. For algebraic simplicity take m odd, with $m = 2m^* + 1$, so that

$$\hat{f}(\omega_p) = \frac{1}{m} \sum_{j=-m^*}^{m^*} I(\omega_p + \frac{2\pi j}{N}).$$

6. Evaluate the degrees of freedom for the Tukey window using equation (7.26).

CHAPTER 8

Bivariate Processes

Thus far, we have been concerned with analysing a single time series. We now turn our attention to the situation where we have observations on *two* time-series and we are interested in the relationship between them.

Jenkins and Watts (1968, p. 322) distinguish two types of situation. In the first type of situation, the two series arise 'on an equal footing' and we are interested in the correlation between them. For example it is often of interest to analyse seismic signals received at two recording sites. In the second, more important type of situation, the two series are 'causally related' and one series is regarded as the *input* to a *linear system*, while the other series is regarded as the *output*. We are then interested in finding the properties of the linear system. The two types of situation are roughly speaking the time series analogues of *correlation* and *regression*.

The first type of situation is considered in this chapter where the cross-correlation function and the cross-spectrum are introduced. These functions are also useful in the study of linear systems which are discussed in Chapter 9.

8.1 Cross-covariance and cross-correlation functions

Suppose we have N observations on two variables, x and y, at unit time intervals over the same period. The observations will be denoted by $(x_1, y_1), \ldots, (x_N, y_N)$. These observations may be regarded as a finite realization of a discrete bivariate stochastic process (X_t, Y_t).

In order to describe a bivariate process it is useful to know the moments up to second order. For a univariate process, the moments up to second order are the mean and autocovariance function. For a bivariate process, the moments up to second order consist of the mean and autovariance functions for each of the two components plus a new function, called the *cross-covariance function*, which is given by

$$\gamma_{xy}(t, \tau) = \text{Cov}(X_t, Y_{t+\tau}).$$

We will only consider bivariate processes which are second-order stationary, so that all moments up to second order do not change with time. We will use the following notation

$$E(X_t) = \mu_x. \quad E(Y_t) = \mu_y.$$
$$\text{Cov}(X_t, X_{t+\tau}) = \gamma_{xx}(\tau)$$
$$\text{Cov}(Y_t, Y_{t+\tau}) = \gamma_{yy}(\tau)$$
$$\text{Cov}(X_t, Y_{t+\tau}) = \gamma_{xy}(\tau) \quad\quad\quad (8.1)$$

Note that some authors (e.g. Granger and Hatanaka, 1964) define the cross-covariance function by

$$\text{Cov}(X_t, Y_{t-\tau}) = \gamma_{xy}^*(\tau).$$

Comparing with (8.1) we see that

$$\gamma_{xy}(\tau) = \gamma_{xy}^*(-\tau).$$

The cross-covariance function differs from the autocovarianc function in that it is *not* an even function, since

$$\gamma_{xy}(\tau) \neq \gamma_{xy}(-\tau).$$

Instead we have the relationship

$$\gamma_{xy}(\tau) = \gamma_{yx}(-\tau)$$

where the subscripts are reversed.

BIVARIATE PROCESSES

The size of the cross-covariance coefficients depends on the units in which $X(t)$ and $Y(t)$ are measured. Thus for interpretative purposes, it is useful to standardize the cross-covariance function to produce a function called the cross-correlation function, $\rho_{xy}(\tau)$, which is defined by

$$\rho_{xy}(\tau) = \gamma_{xy}(\tau)/\sqrt{[\gamma_{xx}(0)\gamma_{yy}(0)]} \tag{8.2}$$

This function measures the correlation between $X(t)$ and $Y(t + \tau)$ and has the properties

(a) $\rho_{xy}(\tau) = \rho_{yx}(-\tau)$
(b) $|\rho_{xy}(\tau)| \leqslant 1$ (see exercise 2).

Whereas $\rho_{xx}(0)$, $\rho_{yy}(0)$ are both equal to one, the value of $\rho_{xy}(0)$ is *not* necessarily equal to one, a fact which is sometimes overlooked.

8.1.1 Examples
Before discussing the estimation of cross-covariance and cross-correlation functions, we will derive the theoretical functions for two examples of bivariate processes. The first example is rather 'artificial', but the model in example 2 can be useful in practice.

Example I   Suppose that $\{X_t\}$, $\{Y_t\}$, are both formed from the same purely random process $\{Z_t\}$ which has mean zero, variance σ_Z^2, by

$$X_t = Z_t$$
$$Y_t = 0.5\, Z_{t-1} + 0.5\, Z_{t-2}.$$

Then using 8.1 we have

$$\gamma_{xy}(\tau) = \begin{cases} 0.5\,\sigma_Z^2 & \tau = 1, 2 \\ 0 & \text{otherwise} \end{cases}$$

Now the variances of the two components are given by

$$\gamma_{xx}(0) = \sigma_Z^2$$

and

$$\gamma_{yy}(0) = \sigma_Z^2/2$$

so that, using (8.2), we have

$$\rho_{xy}(\tau) = \begin{cases} 0.5\sqrt{2} & \tau = 1, 2 \\ 0 & \text{otherwise.} \end{cases}$$

Example 2. Suppose that

$$X_t = Z_{1,t}$$
$$Y_t = X_{t-d} + Z_{2,t} \tag{8.3}$$

where $\{Z_{1,t}\}$, $\{Z_{2,t}\}$ are uncorrelated purely random processes with mean zero and variance σ_Z^2, and where d is an integer. Then we find

$$\gamma_{xy}(\tau) = \begin{cases} \sigma_Z^2 & \tau = d \\ 0 & \text{otherwise} \end{cases}$$

$$\rho_{xy}(\tau) = \begin{cases} 1/\sqrt{2} & \tau = d \\ 0 & \text{otherwise.} \end{cases}$$

In Chapter 9 we will see that equation (8.3) corresponds to putting noise into a linear system which consists of a simple delay of lag d and then adding more noise. The cross-correlation function has a peak at lag d corresponding to the delay in the system, a result which the reader should find intuitively reasonable.

8.1.2 Estimation

The 'obvious' way of estimating the cross-covariance and cross-correlation functions is by means of corresponding sample functions. With N pairs of observations $\{(x_i, y_i);$

172

$i = 1$ to N}, the sample cross-covariance function is

$$c_{xy}(k) = \begin{cases} \sum\limits_{t=1}^{N-k} (x_t - \bar{x})(y_{t+k} - \bar{y})/N & [k = 0, 1, \ldots (N-1)] \\ \sum\limits_{t=1-k}^{N} (x_t - \bar{x})(y_{t+k} - \bar{y})/N \\ & [k = -1, -2, \ldots, -(N-1)] \end{cases} \quad (8.4)$$

and the sample cross-correlation function is

$$r_{xy}(k) = c_{xy}(k)/\sqrt{[c_{xx}(0)c_{yy}(0)]} \quad (8.5)$$

where $c_{xx}(0)$, $c_{yy}(0)$ are the sample variances of observations on x_t and y_t respectively.

It can be shown that these estimators are asymptotically unbiased and consistent. However it can also be shown that successive estimates are themselves autocorrelated. In addition the variances of the estimators depend on the autocorrelation functions of the two components. Thus for moderately large values of N (e.g. N about 100) it is possible for two series, which are actually uncorrelated, to give rise to apparently 'large' cross-correlation coefficients which are actually spurious. Thus if a test is required for non-zero correlation between two time series, both series should first be filtered to convert them to white noise before computing the cross-correlation function (Jenkins and Watts, 1968, p. 340). For two uncorrelated series of white noise it can be shown that

$$E(r_{xy}(k)) \simeq 0$$
$$\text{Var}(r_{xy}(k)) \simeq 1/N$$

so that values outside the interval $\pm 2/\sqrt{N}$ are significantly different from zero.

The filtering procedure mentioned above is accomplished by treating each series separately and fitting an appropriate model. The new filtered series consists of the residuals from this model. For example suppose that one series appeared to be a first-order AR process with estimated parameter $\hat{\alpha}$. Then the filtered series is given by

$$x_t' = (x_t - \bar{x}) - \hat{\alpha}(x_{t-1} - \bar{x}).$$

8.1.3 Interpretation

The interpretation of the sample cross-correlation function can be fraught with danger unless one uses the prefiltering procedure described in Section 8.1.2. For example Coen *et al.* (1969) calculated cross-correlation functions between variables, such as (detrended) Financial Times (F.T.) share index and (detrended) U.K. car production and this resulted in a fairly smooth, roughly sinusoidal function with 'large' coefficients at lags 5 and 6. Coen *et al.* used this information to set up a regression model to 'explain' the variation in the F.T. share index. However Box and Newbold (1971) have shown that the 'large' cross-correlation coefficients are in fact quite spurious as the two series had not been properly filtered. By simply detrending the raw data, Coen *et al.* effectively assumed that the appropriate model for each series was of the form

$$x_t = \alpha + \beta t + a_t$$

where the a_t are *independent*. In fact the error structure was quite different.

If both series are properly filtered, we have seen that it is easy to test whether any of the cross-correlation coefficients are significantly different from zero. Following example 2 of Section 8.1.1, a peak in the estimated cross-correlation function at lag d may indicate that one series is related to the other when delayed by time d.

8.2 The cross-spectrum

The cross-correlation function is the natural tool for examining the relationship between two time series in the

174

time domain. In this section we introduce a complementary function, called the cross spectral density function or cross-spectrum, which is the natural tool in the frequency domain.

By analogy with equation (6.11), we will define the cross-spectrum of a discrete bivariate process measured at unit intervals of time as the Fourier transform of the cross-covariance function, namely

$$f_{xy}(\omega) = \frac{1}{\pi}\left[\sum_{k=-\infty}^{\infty} \gamma_{xy}(k)e^{-i\omega k} \right] \tag{8.6}$$

over the range $0 < \omega < \pi$. The cross-spectrum is sometimes called the cross-power spectrum, although its physical interpretation in terms of power is more difficult than for the autospectrum (see Lee, 1966, p. 348). Indeed a physical understanding of cross-spectra will probably not become clear until we have studied linear systems.

Note that $f_{xy}(\omega)$ is a *complex* function unlike the autospectrum which is real. This is because $\gamma_{xy}(k)$ is not an even function.

The reader must be told at this point that most authors define the cross-spectrum in the range $(-\pi, \pi)$ by analogy with equation (6.13) as

$$f_{xy}(\omega) = \frac{1}{2\pi}\left[\sum_{k=-\infty}^{\infty} \gamma_{xy}(k)e^{-i\omega k} \right] \tag{8.7}$$

This definition has certain mathematical advantages, notably that it can handle complex-valued processes and that it has a simple inverse relationship of the form

$$\gamma_{xy}(k) = \int_{-\pi}^{\pi} e^{i\omega k} f_{xy}(\omega)d\omega \tag{8.8}$$

whereas (8.6) does not have a simple inverse relationship. But (8.7) introduces negative frequencies, and for ease of understanding we shall use (8.6). As regards definition (8.7),

175

note that $f_{xy}(-\omega)$ is the complex conjugate of $f_{xy}(\omega)$ and so provides no extra information. Authors who use (8.7) only examine $f_{xy}(\omega)$ at positive frequencies. Note that Kendall and Stuart (1966) use a different definition to other authors by omitting the constant $1/\pi$ and transforming the cross-correlation function rather than the cross-covariance function.

We now describe several functions derived from the cross-spectrum which are helpful in interpreting the cross-spectrum. From (8.6), the *real* part of the cross-spectrum, called the *co-spectrum* is given by

$$c(\omega) = \frac{1}{\pi} \left[\sum_{k=-\infty}^{\infty} \gamma_{xy}(k) \cos \omega k \right] \qquad (8.9)$$

$$= \frac{1}{\pi} \left\{ \gamma_{xy}(0) + \sum_{k=1}^{\infty} \left[\gamma_{xy}(k) + \gamma_{yx}(k) \right] \cos \omega k \right\}.$$

Note that the formula given by Granger and Hatanaka (1964, p. 75) is incorrect.

The complex part of the cross-spectrum, with a minus sign, is called the *quadrature* spectrum and is given by

$$q(\omega) = \frac{1}{\pi} \left[\sum_{k=-\infty}^{\infty} \gamma_{xy}(k) \sin \omega k \right] \qquad (8.10)$$

$$= \frac{1}{\pi} \left\{ \sum_{k=1}^{\infty} \left[\gamma_{xy}(k) - \gamma_{yx}(k) \right] \sin \omega k \right\}$$

so that

$$f_{xy}(\omega) = c(\omega) - iq(\omega). \qquad (8.11)$$

Note that Granger and Hatanaka (1964, p. 74) express the cross-spectrum as $c(\omega) + iq(\omega)$ in view of their alternative definition of the cross-covariance function.

An alternative way of expressing the cross-spectrum is in the form

$$f_{xy}(\omega) = \alpha_{xy}(\omega) e^{i\phi_{xy}(\omega)} \qquad (8.12)$$

where $\alpha_{xy}(\omega)$ = cross-amplitude spectrum

$$= \sqrt{[c^2(\omega) + q^2(\omega)]} \qquad (8.13)$$

and $\phi_{xy}(\omega)$ = phase spectrum

$$= \tan^{-1}[-q(\omega)/c(\omega)] \qquad (8.14)$$

From (8.14) it appears that $\phi_{xy}(\omega)$ is undetermined by a multiple of π. However if the cross-amplitude spectrum is required to be positive so that we take the positive square root in (8.13) then it can be seen that the phase is actually undetermined by a multiple of 2π using the equality of (8.11) and (8.12). This apparent non-uniqueness makes it difficult to graph the phase. However, when we consider linear systems in chapter 9, we will see that there are physical reasons why the phase *is* generally uniquely determined and not confined to the range $\pm\pi$ or $\pm\pi/2$. The phase is usually zero at $\omega = 0$ and is a continuous function as ω goes from 0 to π.

Another useful function derived from the cross-spectrum is the (squared) coherency which is given by

$$C(\omega) = [c^2(\omega) + q^2(\omega)]/[f_x(\omega)f_y(\omega)] \qquad (8.15)$$
$$= \alpha_{xy}^2(\omega)/f_x(\omega)f_y(\omega)$$

where $f_x(\omega), f_y(\omega)$ are the power spectra of the individual processes, X_t and Y_t. It can be shown that

$$0 \leqslant C(\omega) \leqslant 1.$$

This quantity measures the linear correlation between the two components of the bivariate process at frequency ω and is analogous to the square of the usual correlation coefficient. The closer $C(\omega)$ is to one, the more closely related are the two processes at frequency ω.

Finally we will define a function called the gain spectrum which is given by

$$G_{xy}(\omega) = \sqrt{[f_y(\omega)C(\omega)/f_x(\omega)]}$$
$$= \alpha_{xy}(\omega)/f_x(\omega) \qquad (8.16)$$

177

which is essentially the regression coefficient of the process Y_t on the process X_t at frequency ω. A second gain function can also be defined by $G_{yx}(\omega) = \alpha_{xy}(\omega)/f_y(\omega)$ in which, using linear system terminology, Y_t is regarded as the input and X_t as the output. Note that Granger and Hatanaka (1964) reverse the order of the subscripts of $G_{xy}(\omega)$. Perhaps the notation of Parzen (1964) is clearer, using $G_{y|x}(\omega)$ for expression (8.16).

By this point, the reader will probably be rather confused by all the different functions which have been introduced in relation to the cross-spectrum. Whereas the cross-correlation function is a relatively straightforward development from the autocorrelation function, statisticians often find the cross-spectrum much harder to understand than the autospectrum. Usually *three* functions have to be plotted against frequency to describe the relationship between two series in the frequency domain. Sometimes the co-, quadrature and coherency spectra are most suitable. Sometimes the coherency, phase and cross-amplitude are more appropriate, while another possible trio is coherency, phase and gain. The physical interpretation of these functions will probably not become clear until we have studied linear systems in Chapter 9.

8.2.1 Examples
In this section, we derive the cross-spectrum and related functions for the two examples discussed in Section 8.1.1.

Example 1   Using (8.6) the cross-spectrum is given by

$$f_{xy}(\omega) = [0.5\sigma_Z^2 e^{-i\omega} + 0.5\sigma_Z^2 e^{-2i\omega}]/\pi.$$

Using (8.9), the co-spectrum is given by

$$c(\omega) = 0.5\sigma_Z^2 [\cos \omega + \cos 2\omega]/\pi.$$

Using (8.10), the quadrature spectrum is given by

$$q(\omega) = 0.5\sigma_Z^2 [\sin \omega + \sin 2\omega]/\pi.$$

Using (8.13), the cross-amplitude spectrum is given by

$$\alpha_{xy}(\omega) = \frac{0.5\sigma_Z^2}{\pi}\sqrt{[(\cos \omega + \cos 2\omega)^2 + (\sin \omega + \sin 2\omega)^2]}$$

which, after some algebra, gives

$$\alpha_{xy}(\omega) = \sigma_Z^2 \cos(\omega/2)/\pi.$$

Using (8.14), the phase spectrum is given by

$$\tan \phi_{xy}(\omega) = -(\sin \omega + \sin 2\omega)/(\cos \omega + \cos 2\omega).$$

In order to evaluate the coherency, we need to find the power spectra of the two processes. Since $X_t = Z_t$, it has a constant spectrum given by $f_x(\omega) = \sigma_Z^2/\pi$. The spectrum of Y_t is given by

$$f_y(\omega) = 0.5\sigma_Z^2(1 + \cos \omega)/\pi$$
$$= \sigma_Z^2 \cos^2(\omega/2)/\pi.$$

Thus, using (8.15), the coherency spectrum is given by

$$C(\omega) = 1 \qquad \text{for all } \omega \text{ in } (0, \pi).$$

This latter result may at first sight appear surprising. But both X_t and Y_t are generated from the *same* noise process and this explains why there is perfect correlation between the components of the two processes at any given frequency. Finally, using (8.16), the gain spectrum is given by

$$G_{xy}(\omega) = \cos(\omega/2).$$

Since the coherency is unity, $G_{xy}(\omega) = \sqrt{[f_y(\omega)/f_x(\omega)]}$.

Example 2   In this case we find

$$f_{xy}(\omega) = \sigma_Z^2 e^{-i\omega d}/\pi$$
$$c(\omega) = \sigma_Z^2 \cos \omega d/\pi$$
$$q(\omega) = \sigma_Z^2 \sin \omega d/\pi$$
$$\alpha_{xy}(\omega) = \sigma_Z^2/\pi$$
$$\tan \phi_{xy}(\omega) = -\tan \omega d. \tag{8.17}$$

179

Then, as the two autospectra are given by

$$f_x(\omega) = \sigma_Z^2/\pi$$

and

$$f_y(\omega) = 2\sigma_Z^2/\pi$$

we find

$$C(\omega) = 1/2.$$

The function of particular interest in this example is the phase, which from (8.17), is a straight line with slope $(-d)$ when $\phi_{xy}(\omega)$ is unconstrained and is plotted against ω as a continuous function starting with zero phase at zero frequency (see Fig. 8.1(a)). If, however, the phase is constrained to lie within the interval $(-\pi, \pi)$ then a graph like Fig. 8.1(b) will result where the slope of each line is $(-d)$.

This result is often used in identifying relationships between time series. If the estimated phase approximates a straight line through the origin then this indicates a delay between the two series equal to the slope of the line. An example of this in practice is given by Barksdale and Guffey (1972). More generally, the time delay between two recording sites will change with frequency, due, for example, to varying speeds of propagation. This is called the *dispersive* case and real-life examples are discussed by Haubrich and Mackenzie (1965) and Hamon and Hannan (1974).

8.2.2 Estimation

As in Section 7.4, there are two basic approaches to estimating the cross-spectrum. Firstly we can take a Fourier transform of the truncated sample cross-covariance function (or of the cross-correlation function to get a normalized cross-spectrum). The estimated co-spectrum is then given by

$$\hat{c}(\omega) = \frac{1}{\pi} \left[\sum_{k=-M}^{M} \lambda_k c_{xy}(k) \cos \omega k \right] \tag{8.18}$$

180

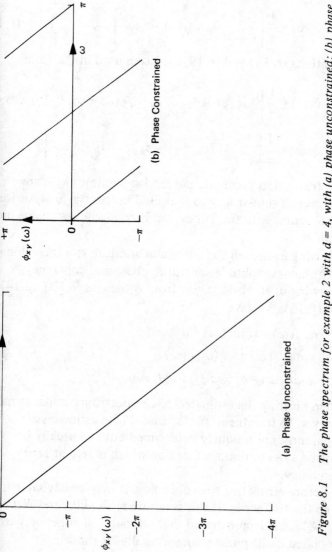

Figure 8.1   The phase spectrum for example 2 with $d = 4$, with (a) phase unconstrained; (b) phase constrained.

where M is the truncation point, and $\{\lambda_k\}$ is the lag window. The estimated quadrature spectrum is given by

$$\hat{q}(\omega) = \frac{1}{\pi} \left[\sum_{k=-M}^{M} \lambda_k c_{xy}(k) \sin \omega k \right] \qquad (8.1$$

Equations (8.18) and (8.19) are often used in the form

$$\hat{c}(\omega) = \frac{1}{\pi} \left\{ \lambda_0 c_{xy}(0) + \sum_{k=1}^{M} \lambda_k [c_{xy}(k) + c_{xy}(-k)] \cos \right.$$

$$\hat{q}(\omega) = \frac{1}{\pi} \left\{ \sum_{k=1}^{M} \lambda_k [c_{xy}(k) - c_{xy}(-k)] \sin \omega k \right\}$$

The truncation point, M, and the lag window $\{\lambda_k\}$, are chosen in a similar way to that used in spectral analysis for single series, with the Tukey and Parzen windows being mc popular.

Having estimated the co- and quadrature spectra, estima of the cross-amplitude spectrum, phase and coherency following in an obvious way from equations (8.13), (8.14) and (8.15). We have

$$\hat{\alpha}_{xy}(\omega) = \sqrt{[\hat{c}^2(\omega) + \hat{q}^2(\omega)]}$$

$$\tan \hat{\phi}_{xy}(\omega) = -\hat{q}(\omega)/\hat{c}(\omega)$$

$$\text{and } \hat{C}(\omega) = \hat{\alpha}_{xy}^2(\omega)/\hat{f}_x(\omega)\hat{f}_y(\omega).$$

When plotting the estimated phase spectrum, similar remark apply as to the (theoretical) phase. Phase estimates are apparently not uniquely determined but can usually be plotted as a continuous function which is zero at zero frequency.

Before estimating the coherency, it may be advisable to *align* the two series. If this is not done, Jenkins and Watts (1968) have demonstrated that estimates of coherency wil be biased if the phase changes rapidly. If the sample cross-correlation function has its largest value at lag s, the

182

ᴡe two series are aligned by translating one of the series a
istance s so that the peak in the cross-correlation function
f the aligned series is at zero lag.

The second approach to cross-spectral analysis is to
ᴍooth a function called the *cross-periodogram*. The
nivariate periodogram of a series $\{x_t\}$ can be written in the
ᴏrm

$$I(\omega_p) = (\Sigma x_t\, e^{i\omega_p t})(\Sigma x_t\, e^{-i\omega_p t})/N\pi \qquad (8.20)$$
$$= N(a_p^2 + b_p^2)/4\pi$$

ᴜsing (7.17) and (7.10). By analogy with (8.20) we may
ᴇfine the cross-periodogram of two series (x_t) and (y_t) as

$$I_{xy}(\omega_p) = (\Sigma x_t\, e^{i\omega_p t})(\Sigma y_t\, e^{-i\omega_p t})/N\pi \qquad (8.21)$$

ᴠe then find that the real and imaginary parts of $I_{xy}(\omega_p)$
ᴀre given by

$$N(a_{px}a_{py} + b_{px}b_{py})/4\pi \text{ and } N(a_{px}b_{py} - a_{py}b_{px})/4\pi$$

ᴠhere (a_{px}, b_{px}), (a_{py}, b_{py}) are the Fourier coefficients of
$\{x_t\}$, $\{y_t\}$ at ω_p. These real and imaginary parts may then
ᴇe smoothed to get consistent estimates of the co- and
ᴜuadrature spectral density functions by

$$\hat{c}(\omega_p) = N \sum_{q=p-m^*}^{p+m^*} (a_{qx}a_{qy} + b_{qx}b_{qy})/4\pi m$$

$$\hat{q}(\omega_p) = N \sum_{q=p-m^*}^{p+m^*} (a_{qx}b_{qy} - a_{qy}b_{qx})/4\pi m$$

where $m = 2m^* + 1$. These estimates may then be used to
estimate the cross-amplitude spectrum, phase, etc. as before.

This type of approach has been rather neglected in the
ᴌiterature yet the computational advantages are clear. Once a
periodogram analysis has been made of the two individual
processes, nearly all the work has been done as the estimates

of $c(\omega)$ and $q(\omega)$ only involve the Fourier coefficients of th two series. The disadvantage of the approach is that alignment is only possible if the cross-correlation function is calculated separately. This can be done directly or by the us of two (fast) Fourier transforms by an analogous procedure to that described in Section 7.4.5. Tick (1967), suggests several smoothing procedures as alternatives to alignment. Further work needs to be done on this approach, but it seem likely to displace the traditional method for long series.

We will not say much about the properties of cross-spectr estimators. Asymptotic results are given by Jenkins and Watt (1968) while Granger and Hughes (1968) have carried out a simulation study on some short series. The following points are worth noting. Estimates of phase and cross-amplitude are imprecise when the coherency is relatively small. Estimates o coherency are constrained to lie between 0 and 1, and there may be a bias towards ½ which may be serious with short series. Rapid changes in phase may bias coherency estimates. Generally speaking, this is·an area of time-series analysis where much work remains to be done.

8.2.3 Interpretation

Cross-spectral analysis is a technique for examining the relationship between two series over a range of frequencies. The technique may be used for two time series which 'arise on a similar footing' and then the coherency spectrum is perhaps the most useful function. It measures the linear correlation between two series at each frequency and is analogous to the square of the ordinary product—moment correlation coefficient.

The other functions introduced in this chapter, such as the phase spectrum, are more readily understood in the context of linear systems which will be discussed in Chapter 9. We wil therefore defer further discussion of how to interpret cross-spectral estimates until Section 9.3.

Exercises

1. Show that the cross-covariance function of the discrete bivariate process (X_t, Y_t) where

$$X_t = Z_{1,t} + \beta_{11} Z_{1,t-1} + \beta_{12} Z_{2,t-1}$$
$$Y_t = Z_{2,t} + \beta_{21} Z_{1,t-1} + \beta_{22} Z_{2,t-1}$$

and $(Z_{1,t})$, $(Z_{2,t})$ are independent purely random processes with zero mean and variance σ_Z^2, is given by

$$\gamma_{xy}(k) = \begin{cases} \sigma_Z^2 (\beta_{11}\beta_{21} + \beta_{12}\beta_{22}) & k = 0 \\ \beta_{21} \sigma_Z^2 & k = 1 \\ \beta_{12} \sigma_Z^2 & k = -1 \\ 0 & \text{otherwise.} \end{cases}$$

Hence evaluate the cross-spectrum.

2. Define the cross-correlation function, $\rho_{xy}(\tau)$, of a bivariate stationary process and show that $|\rho_{xy}(\tau)| \leqslant 1$ for all τ. Two MA processes

$$X_t = Z_t + 0.4\, Z_{t-1}$$
$$Y_t = Z_t - 0.4\, Z_{t-1}$$

are formed from a purely random process, $\{Z_t\}$, which has mean zero and variance σ_Z^2. Find the cross-covariance and cross-correlation functions of the bivariate process $\{X_t, Y_t\}$ and hence show that the cross-spectrum is given by

$$f_{xy}(\omega) = \sigma_Z^2 (0.84 + 0.8\, i \sin \omega)/\pi \quad (0 < \omega < \pi).$$

Evaluate the co-, quadrature, cross-amplitude, phase and coherency spectra.

CHAPTER 9

Linear Systems

9.1 Introduction

An important problem in engineering and the physical sciences is that of *identifying* a model for a physical system (or process) given observations on the input and output to the system. For example, the yield from a chemical reactor (the output) depends *inter alia* on the temperature at which the reactor is kept (the input). Much of the literature assume that the system can be adequately approximated over the range of interest by a *linear* model whose parameters do not change with time, though recently there has been increased interest in time-varying and non-linear systems. Some general references include Astrom and Eykhoff (1971), Eykhoff (1974) and Harris (1976).

We shall denote the input and output series by $\{x_t\}, \{y_t\}$ respectively in discrete time, and by $\{x(t)\}, \{y(t)\}$ respectively in continuous time.

In this chapter we confine attention to linear systems (e.g. Schwarz and Friedland, 1965). A precise definition of linearity is as follows. Suppose $y_1(t), y_2(t)$ are the outputs corresponding to $x_1(t), x_2(t)$ respectively. Then the system is said to be linear if, and only if, a linear combination of the inputs, say $\lambda_1 x_1(t) + \lambda_2 x_2(t)$, produces the same linear combination of the outputs, namely $\lambda_1 y_1(t) + \lambda_2 y_2(t)$, where λ_1, λ_2 are any constants.

We shall further confine our attention to linear systems which are *time-invariant*. This term is defined as follows. If input $x(t)$ produces output $y(t)$, then the system is said to be time-invariant if a delay of time τ in the input produces the same delay in the output. In other words $x(t - \tau)$ produces output $y(t - \tau)$, so that the input–output relation does not change with time.

We will only consider systems having one input and one output. The extension to several inputs and outputs is straightforward in principle, though difficult in practice.

The study of linear systems is not only useful for the purpose of examining the relationship between different time series, but also for examining the properties of filtering procedures such as detrending.

In Sections 9.2 and 9.3, we show how to describe linear systems in the time and frequency domains respectively, while in Section 9.4 we discuss the identification of linear systems from observed data.

9.2 Linear systems in the time domain

A time-invariant linear system may generally be written in the form

$$y(t) = \int_{-\infty}^{\infty} h(u)x(t - u)du \qquad (9.1)$$

in continuous time, or

$$y_t = \sum_{k = -\infty}^{\infty} h_k x_{t - k} \qquad (9.2)$$

in discrete time. The weight function, $h(u)$ in continuous time or $\{h_k\}$ in discrete time, provides a description of the system in the time domain. This function is called the *impulse response function* of the system for reasons which will become apparent later. Note that some engineers prefer to call the discrete function, $\{h_k\}$, the *unit sample response function*.

It is clear that equations (9.1) and (9.2) are linear. The property of time-invariance ensures that the impulse response function does not depend on t. The system is said to be *physically realizable* or causal if

$$h(u) = 0 \qquad (u < 0)$$

or

$$h_k = 0 \qquad (k < 0).$$

Engineers have been principally concerned with continuous-time systems but are increasingly studying sampled-data control problems. Statisticians generally work with discrete data and so the subsequent discussion is mainly concerned with the discrete case.

We will only consider *stable* systems for which a bounded input produces a bounded output, although control engineers are frequently concerned with the control of unstable systems. A sufficient condition for stability is that the impulse response function should satisfy

$$\sum_k |h_k| < C$$

where C is a finite constant.

9.2.1 Some examples

The linear filters introduced in Section 2.5.2 are examples of linear systems. For example the simple moving average given by

$$y_t = (x_{t-1} + x_t + x_{t+1})/3$$

has impulse response function

$$h_k = \begin{cases} \dfrac{1}{3} & k = -1, 0, +1 \\ 0 & \text{otherwise} \end{cases}$$

Note that this filter is not 'physically realizable', although it can of course be used as a mathematical smoothing device.

Another general class of linear systems are those expressed as linear differential equations with constant coefficients in continuous time. For example, if T is a constant, then

$$T \frac{dy(t)}{dt} + y(t) = x(t).$$

is a description of a linear system. In discrete time, the analogue of differential equations are *difference* equations given by

$$y_t + \alpha_1 \nabla y_t + \alpha_2 \nabla^2 y_t + \ldots$$
$$= \beta_0 x_t + \beta_1 \nabla x_t + \beta_2 \nabla^2 x_t + \ldots \quad (9.3)$$

where $\nabla y_t = y_t - y_{t-1}$. Equation (9.3) can be rewritten as

$$y_t = a_1 y_{t-1} + a_2 y_{t-2} + \ldots$$
$$+ b_0 x_t + b_1 x_{t-1} + \ldots \quad (9.4)$$

It is clear that equation (9.4) can be rewritten in the form (9.2) by successive substitution. For example if

$$y_t = \tfrac{1}{2} y_{t-1} + x_t$$

then

$$y_t = x_t + \tfrac{1}{2} x_{t-1} + \tfrac{1}{4} x_{t-2} \ldots$$

so that the impulse response function is given by

$$h_k = \begin{cases} (\tfrac{1}{2})^k & k = 0, 1, \ldots \\ 0 & k < 0 \end{cases}$$

Two very simple linear systems are given by

$$y_t = x_{t-d} \quad (9.5)$$

called *simple delay*, where the integer d denotes the delay time time, and

$$y_t = g x_t \quad (9.6)$$

called *simple gain*, where g is a constant called the gain. The impulse response functions of (9.5) and (9.6) are

$$h_k = \begin{cases} 1 & k = d \\ 0 & \text{otherwise} \end{cases}$$

and

$$h_k = \begin{cases} g & k = 0 \\ 0 & \text{otherwise} \end{cases}$$

respectively.

In continuous time, the impulse response functions of simple delay and simple gain, namely

$$y(t) = x(t - \tau)$$

and

$$y(t) = gx(t)$$

can only be represented in terms of the Dirac delta function (see Appendix II). The functions are $\delta(u - \tau)$ and $g\delta(u)$ respectively.

An important class of impulse response functions, which often provides a reasonable approximation to physically realizable systems, is given by

$$h(u) = \begin{cases} [ge^{-(u-\tau)/T}]/T & (u > \tau) \\ 0 & (u < \tau) \end{cases}$$

A function of this type is called a *delayed exponential*, and depends on three constants, g, T, and τ. The constant τ is called the *delay*. When $\tau = 0$, we have simple exponential response. The constant g is called the *gain*, and represents the eventual change in output when a step change of unit size is made to the input. The constant T governs the rate at which the output changes. Fig. (9.1) shows how the output to a delayed exponential system changes when a step change of unity is made to the input.

190

(a) graph of impulse response function

(b) graph of input

(c) graph of output

Figure 9.1   A delayed exponential response to a unit step change in input.

9.2.2 The impulse response function The impulse response function describes how the output is related to the input of a linear system. (See equations 9.1 and 9.2.) The name 'impulse response' arises from the fact that the function describes the response of the system to an impulse input. For example, in discrete time, suppose that the input, x_t, is zero for all t except at time zero when it takes the value unity, so that $x_0 = 1$. Then the output at time t is given by

$$y_t = \Sigma h_k x_{t-k}$$
$$= h_t$$

Thus the output resulting from the unit impulse input is the same as the impulse response function.

9.2.3 The step response function An alternative, equivalent, way of describing a linear system in the time domain is by means of a function called the step response function, which is defined by

$$S(t) = \int_{-\infty}^{t} h(u)\mathrm{d}u \tag{9.7}$$

in continuous time, and

$$S_t = \sum_{k \leqslant t} h_k \tag{9.8}$$

in discrete time.

The name 'step response' arises from the fact that the function describes the response of the system to a unit step change in the input. For example, in discrete time, suppose that the input is given by

$$x_t = \begin{cases} 0 & t < 0 \\ 1 & t \geqslant 0 \end{cases}$$

then $y_t = \sum\limits_k h_k x_{t-k}$

$\qquad = \sum\limits_{k \leq t} h_k$

$\qquad = S_t,$

so that the output is equal to the step response function.

Engineers sometimes use this relationship to measure the properties of a physically realizable system. The input is held steady for some time and then a unit step change is made to the input. The output is then observed and this provides an estimate of the step response function, and hence, of its derivative, the impulse response function. A step change in the input may be easier to provide than an impulse.

The step response function for a delayed exponential system is given by

$$S(t) = g[1 - e^{-(t-\tau)/T}] \quad (t > \tau) \qquad (9.9)$$

and the graph of $y(t)$ in Fig. (9.1) is also a graph of $S(t)$.

9.3 Linear systems in the frequency domain

9.3.1 The frequency response function
An alternative way of describing a linear system is by means of a function, called the *frequency response function* or *transfer function*, which is the Fourier transform of the impulse response function. It is defined by

$$H(\omega) = \int_{-\infty}^{\infty} h(u)e^{-i\omega u} du \qquad (0 < \omega < \infty) \qquad (9.10)$$

in continuous time, and

$$H(\omega) = \sum\limits_k h_k e^{-i\omega k} \qquad (0 < \omega < \pi) \qquad (9.11)$$

in discrete time.

The frequency response and impulse response functions are equivalent ways of describing a linear system, in a somewhat similar way that the autocovariance and spectral density functions are equivalent ways of describing a stationary stochastic process; one function being the Fourier transform of the other. We shall see that, for some purposes, $H(\omega)$ is much more useful than $h(u)$. First we prove the following theorem.

THEOREM 9.1   A sinusoidal input to a linear system gives rise, in the steady-state, to a sinusoidal output at the *same* frequency. The amplitude of the sinusoid may change and there may also be a phase shift.

Proof   The proof is given for continuous time, the extension to discrete time being straightforward. Suppose that the input to a linear system, with impulse response function, $h(u)$, is given by

$$x(t) = \cos \omega t, \quad \text{for all } t.$$

Then the output is given by

$$y(t) = \int_{-\infty}^{\infty} h(u) \cos \omega(t - u) \, du \tag{9.12}$$

Now $\cos(A - B) = \cos A \cos B + \sin A \sin B$, so we may rewrite (9.12) as

$$y(t) = \cos \omega t \int_{-\infty}^{\infty} h(u) \cos \omega u \, du.$$

$$+ \sin \omega t \int_{-\infty}^{\infty} h(u) \sin \omega u \, du.$$

As the two integrals do not depend on t, it is now obvious that $y(t)$ is a mixture of sine and cosine terms at frequency ω. Thus the output is a sinusoidal perturbation at the same frequency, ω, as the input.

194

If we write

$$A(\omega) = \int_{-\infty}^{\infty} h(u) \cos \omega u \, du$$

$$B(\omega) = \int_{-\infty}^{\infty} h(u) \sin \omega u \, du$$

$$G(\omega) = \sqrt{[A^2(\omega) + B^2(\omega)]} \qquad (9.13)$$

and $\tan \phi(\omega) = -B(\omega)/A(\omega)$ $\qquad (9.14)$

then

$$y(t) = A(\omega) \cos \omega t + B(\omega) \sin \omega t$$
$$= G(\omega) \cos(\omega t + \phi(\omega)) \qquad (9.15)$$

Equation (9.15) shows that a cosine wave is amplified by a factor $G(\omega)$, which is called the *gain* of the system. The equation also shows that the cosine wave is shifted by an angle $\phi(\omega)$, which is called the *phase shift*. Note that both the gain and phase shift may vary with frequency. From equation (9.14) we see that the phase shift is apparently not uniquely determined. If we take the positive square root in equation (9.13), so that the gain is required to be positive, then the phase shift is undetermined by a multiple of 2π (see also Sections 8.2 and 9.3.2).

We have so far considered an input cosine wave. By a similar argument it can be shown that an input sine wave, $x(t) = \sin \omega t$, gives an output $y(t) = G(\omega) \sin(\omega t + \phi(\omega))$, so that there is the same gain and phase shift. More generally if we consider an input complex exponential given by

$$x(t) = e^{i\omega t}$$

$$= \cos \omega t + i \sin \omega t$$

then the output is given by

$$y(t) = G(\omega)\{\cos[\omega t + \phi(\omega)] + i \sin[\omega t + \phi(\omega)]\}$$
$$= G(\omega)e^{i[\omega t + \phi(\omega)]}$$
$$= G(\omega)e^{i\phi(\omega)}x(t). \qquad (9.16)$$

195

Now

$$G(\omega)e^{i\phi(\omega)} = A(\omega) - iB(\omega)$$

$$= \int_{-\infty}^{\infty} h(u)(\cos \omega u - i \sin \omega u)du$$

$$= \int_{-\infty}^{\infty} h(u)e^{-i\omega u}\,du$$

$$= H(\omega) \tag{9.17}$$

So when the input in equation (9.16) is of the form $e^{i\omega t}$, the output is given simply by (frequency response function x input) and we have (in the steady-state situation)

$$y(t) = H(\omega)\,x(t) \tag{9.18}$$

This completes the proof of Theorem 9.1.

Transients   The reader should note that theorem 9.1 only applies in the *steady-state* where it is assumed that the input sinusoid was applied at $t = -\infty$. If in fact the sinusoid is applied at say $t = 0$, then the output will take some time to settle to the steady-state form given by the theorem. The difference between the observed output and the steady-state output is called the *transient* component. The system is stable if this transient component tends to zero as $t \to \infty$. If the relationship between input and output is expressed as a differential (or difference) equation, then the steady-state solution corresponds to the particular integral, while the transient component corresponds to the complementary function.

It is easier to describe the transient behaviour of a linear system by using the *Laplace* transform of the impulse response function. Engineers also prefer the Laplace transform as it is defined for unstable systems. However statisticians have customarily dealt with steady-state behaviour and used Fourier transforms, and we will continue

is custom. However this is certainly an aspect of linear
stems which statisticians should look at more closely.

SCUSSION OF THEOREM 9.1   Theorem 9.1 helps to
troduce the importance of the frequency response
nction. For inputs consisting of an impulse or step change
it easy to calculate the output using the impulse response
nction. But for a sinusoidal input, it is much easier to
lculate the output using the frequency response function.
ompare equations (9.12) and (9.18).] More generally for
input consisting of several sinusoidal perturbations,
mely

$$x(t) = \sum_j A_j(\omega_j)e^{i\omega_j t}$$

is easy to calculate the output using the frequency response
nction as

$$y(t) = \sum_j A_j(\omega_j)H(\omega_j)e^{i\omega_j t}.$$

us a complicated convolution in the time domain, as in
uation (9.12), reduces to a simple multiplication in the
quency domain, and we shall see that linear systems are
ten easier to study in the frequency domain using the
quency response function.
Returning to the definition of the frequency response
nction as given by equations (9.10) and (9.11), note that
me authors define $H(\omega)$ for negative as well as positive
quencies. But for real-valued processes we need only
nsider $H(\omega)$ for $\omega > 0$. Note that, in discrete time, $H(\omega)$ is
ly defined for frequencies up to the Nyquist frequency, π
r $\pi/\Delta t$ if there is an interval Δt between successive
servations). We have already introduced the Nyquist
quency in Section 7.2.1. Applying similar ideas to a linear
stem, it is clear that a sinusoidal input which has a higher
quency than π will have a corresponding sinusoid at a
quency in $(0, \pi)$ which gives identical readings at unit

intervals of time and which will therefore give rise to an identical output.

We have already noted that $H(\omega)$ is sometimes called the frequency response function and sometimes the transfer function. We will use the former term as it is more descriptive, indicating that the function shows how a linear system responds to sinusoids at different frequencies. In any case the term transfer function is used by some authors in a different way. Engineers use the term to denote the *Laplace* transform of the impulse response function (see Appendix For a physically realizable stable system, the Fourier transform of the impulse response function may be regarded as a special case of the Laplace transform. A necessary and sufficient condition for a linear system to be stable is that Laplace transform of the impulse response function should have no poles in the right half plane or on the imaginary axis For an unstable system, the Fourier transform does not exist but the Laplace transform does. But we will only consider stable systems, in which case the Fourier transform is adequate. Note that Jenkins and Watts (1968) use the term 'transfer function' to denote the Z-transform of the impulse response function (see Appendix I) in the discrete case. Z-transforms are also used by engineers for discrete-time systems (e.g. Schwarz and Friedland, 1965).

9.3.2 Gain and phase diagrams   The frequency response function, $H(\omega)$, of a linear system is a complex function given by

$$H(\omega) = G(\omega)e^{i\phi(\omega)}$$

where $G(\omega), \phi(\omega)$ are the gain and phase respectively. In order to understand the frequency properties of the system it is useful to plot $G(\omega)$ and $\phi(\omega)$ against ω to obtain what are called the *gain diagram* and the *phase diagram*. If $G(\omega)$ 'large' for low values of ω, but 'small' for high values of ω in Fig. 9.2(a), then we have what is called a *low-pass* filter

198

Figure 9.2   Gain diagrams for (a) a low-pass filter; (b) a high-pass filter; (c) a simple moving average of three successive observations.

This description is self-explanatory in that, if the input is a mixture of variation at several different frequencies, only those components with a low frequency will 'get through' the filter.

Conversely, if $G(\omega)$ is 'small' for low values of ω, but 'large' for high values of ω, then we have a *high-pass* filter as in Fig. 9.2(b).

Plotting the phase diagram is complicated by the fact that the phase in equation (9.14) is not uniquely determined. If the gain is always taken to be positive, then the phase is undetermined by a multiple of 2π and is often constrained to the range $(-\pi, \pi)$. The (complex) value of $H(\omega)$ is examined to see which quadrant it is in. If $G(\omega)$ can be positive or negative, the phase is undetermined by a multiple of π, and is often constrained to the range $(-\pi/2, \pi/2)$. These different conventions are not discussed adequately in many books (Hause, 1971, p. 214). In fact there are physical reasons why engineers prefer to plot the phase as a continuous unconstrained function allowing $G(\omega)$ to be positive or negative, and using the fact that $\phi(0) = 0$ provided $G(0)$ is finite.

9.3.3 Some examples
Example I   Consider the simple moving average

$$y_t = (x_{t-1} + x_t + x_{t+1})/3$$

which is a linear system with impulse response function

$$h_k = \begin{cases} \dfrac{1}{3} & k = -1, 0, +1 \\ 0 & \text{otherwise} \end{cases}$$

The frequency response function of this filter is (using equation 9.11)

$$H(\omega) = \frac{1}{3} e^{-i\omega} + \frac{1}{3} + \frac{1}{3} e^{i\omega}$$

$$= \frac{1}{3} + \frac{2}{3} \cos \omega \qquad (0 < \omega < \pi).$$

This function happens to be real, not complex, and so the phase appears to be given by

$$\phi(\omega) = 0 \qquad (0 < \omega < \pi).$$

However $H(\omega)$ is negative for $\omega > 2\pi/3$, and so if we adopt the convention that the gain should be positive, then we have

$$G(\omega) = \left| \frac{1}{3} + \frac{2}{3} \cos \omega \right|$$

$$= \begin{cases} \dfrac{1}{3} + \dfrac{2}{3} \cos \omega & 0 < \omega < 2\pi/3 \\[2mm] -\dfrac{1}{3} - \dfrac{2}{3} \cos \omega & 2\pi/3 < \omega < \pi \end{cases}$$

and

$$\phi(\omega) = \begin{cases} 0 & 0 < \omega < 2\pi/3 \\ \pi & 2\pi/3 < \omega < \pi. \end{cases}$$

The gain is plotted in Fig. 9.2(c) and is of low-pass type. This is to be expected as a moving average smooths out local fluctuations (high frequency variation) and measures the trend (the low frequency variation). In fact it is probably more sensible to allow the gain to go negative in $(2\pi/3, \pi)$ so that the phase is zero for all ω in $(0, \pi)$.

Example 2 A linear system showing simple exponential response has impulse response function

$$h(u) = g\, e^{-u/T}/T \qquad (u > 0).$$

201

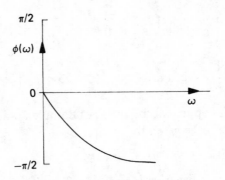

Figure 9.3   Phase diagram for a simple exponential response system.

Using (9.10), the frequency response function is

$$H(\omega) = g(1 - i\omega T)/(1 + \omega^2 T^2) \qquad (\omega > 0).$$

Hence

$$G(\omega) = g/\sqrt{(1 + \omega^2 T^2)},$$

and

$$\tan \phi(\omega) = -T\omega.$$

As the frequency increases, $G(\omega)$ decreases so that the system is of low-pass type. As regards the phase, if we take $\phi(\omega)$ to be zero at zero frequency, then the phase becomes increasingly negative as ω increases until the output is out of phase with the input (see Fig. 9.3).

Example 3   Consider the linear system consisting of pure delay, so that

$$y(t) = x(t - \tau), \text{ where } \tau \text{ is a constant.}$$

Jenkins and Watts (1968) give the impulse response function as

$$h(u) = \delta(u - \tau)$$

where δ denotes the Dirac delta function (see Appendix II).

Then the frequency response function is given by

$$H(\omega) = \int_{-\infty}^{\infty} \delta(u - \tau)e^{-i\omega u}du$$

$$= e^{-i\omega\tau}.$$

In fact $H(\omega)$ can be derived without bringing in delta functions by using theorem 9.1. Suppose that input $x(t) = e^{i\omega t}$ is applied to the system. Then the output is $y(t) = e^{i\omega(t-\tau)} = e^{-i\omega\tau} \times$ input. Thus, by analogy with equation (9.18), we have $H(\omega) = e^{-i\omega\tau}$.

For this linear system, the gain is constant, namely

$$G(\omega) = 1,$$

while the phase is given by

$$\phi(\omega) = -\omega\tau.$$

9.3.4 General relation between input and output   We have, so far, considered only sinusoidal inputs in the frequency domain. In this section we consider any type of input and show that it is generally easier to work with linear systems in the frequency domain, than in the time domain.

The general relation between input and output in the time domain is given by equation (9.1), namely

$$y(t) = \int_{-\infty}^{\infty} h(u)x(t - u)du \tag{9.19}$$

When $x(t)$ is not of a simple form, this integral may be hard to evaluate. Now consider the Fourier transform of the output by

$$Y(\omega) = \int_{-\infty}^{\infty} y(t)e^{-i\omega t}dt$$

$$= \int_{-\infty}^{\infty} \int_{-\infty}^{\infty} h(u)x(t - u)e^{-i\omega t}du\,dt$$

$$= \int_{-\infty}^{\infty} \int_{-\infty}^{\infty} h(u)e^{-i\omega u}\,x(t - u)e^{-i\omega(t - u)}\,du\,dt.$$

But

$$\int_{-\infty}^{\infty} x(t-u)e^{-i\omega(t-u)}\,dt = \int_{-\infty}^{\infty} x(t)e^{-i\omega t}\,dt$$

for all values of u, and is therefore the Fourier transform of $x(t)$ which we will denote by $X(\omega)$. And

$$\int_{-\infty}^{\infty} h(u)e^{-i\omega u}\,du = H(\omega)$$

so that

$$Y(\omega) = H(\omega)X(\omega) \tag{9.20}$$

Thus the integral in (9.19) corresponds to a multiplication in the frequency domain provided that the Fourier transforms exist. A similar result holds in discrete time.

A more useful general relation between input and output, akin to equation (9.20), can be obtained when the input $x(t)$ is a stationary process with continuous power spectrum. This result will be given as theorem 9.2.

THEOREM 9.2   Consider a stable linear system with gain function, $G(\omega)$. Suppose that the input $X(t)$ is a stationary process with continuous power spectrum $f_x(\omega)$. Then the output $Y(t)$ is also a stationary process, whose power spectrum, $f_y(\omega)$, is given by

$$f_y(\omega) = G^2(\omega)f_x(\omega) \tag{9.21}$$

Proof   The proof will be given for continuous time, but the same result holds for discrete time. Let us denote the impulse response and frequency response functions of the system by $h(u)$, $H(\omega)$ respectively. Thus $G(\omega) = |H(\omega)|$.

It is easy to show that a stationary input to a stable linear system gives rise to a stationary output and this will not be shown here. For mathematical convenience, let us assume that the input has mean zero. Then the output

$$Y(t) = \int_{-\infty}^{\infty} h(u)X(t - u)\, du$$

also has mean zero.

Denote the autocovariance functions of $X(t)$, $Y(t)$ by $\gamma_x(\tau)$, $\gamma_y(\tau)$ respectively. Then

$$\gamma_y(\tau) = E[Y(t)Y(t + \tau)] \qquad \text{since} \qquad E[Y(t)] = 0$$

$$= E\left\{ \int_{-\infty}^{\infty} h(u)X(t - u)du \int_{-\infty}^{\infty} h(u')X(t + \tau - u')du' \right\}.$$

$$= \iint_{-\infty}^{\infty} h(u)h(u')E[X(t - u)X(t + \tau - u')]\, du\, du'$$

But

$$E[X(t - u)X(t + \tau - u')] = \gamma_x(\tau - u' + u).$$

Thus

$$\gamma_y(\tau) = \iint_{-\infty}^{\infty} h(u)h(u')\gamma_x(\tau - u' + u)du\, du' \qquad (9.22)$$

The relationship (9.22) between the autocovariance functions of input and output is not of a simple form. However if we take Fourier transforms of both sides of (9.22), by multiplying by $e^{-i\omega\tau}/\pi$ and integrating with respect to τ, we find for the left hand side

$$\frac{1}{\pi} \int_{-\infty}^{\infty} \gamma_y(\tau)e^{-i\omega\tau}\, d\tau = fy(\omega)$$

from equation (6.17), while for the right hand side

$$\iint_{-\infty}^{\infty} h(u)h(u')\left[\frac{1}{\pi} \int_{-\infty}^{\infty} \gamma_x(\tau - u' + u)e^{-i\omega\tau}\, d\tau \right]du\, du'$$

$$= \iint_{-\infty}^{\infty} h(u)e^{i\omega u}h(u')e^{-i\omega u'}$$

$$\times \left[\frac{1}{\pi} \int_{-\infty}^{\infty} \gamma_x(\tau - u' + u)e^{-i\omega(\tau - u' + u)}\, d\tau \right]du\, du'$$

But

$$\frac{1}{\pi} \int_{-\infty}^{\infty} \gamma_x(\tau - u' + u) e^{-i\omega(\tau - u' + u)} d\tau$$

$$= \frac{1}{\pi} \int_{-\infty}^{\infty} \gamma_x(\tau) e^{-i\omega\tau} d\tau = f_x(\omega)$$

for all u, u', and

$$\int_{-\infty}^{\infty} h(u) e^{i\omega u} = \overline{H(\omega)}$$

$$= G(\omega) e^{-i\phi(\omega)}$$

Thus

$$f_y(\omega) = \overline{H(\omega)} \, H(\omega) \, f_x(\omega)$$

$$= G^2(\omega) f_x(\omega)$$

This completes the proof.

The relationship between the spectra of input and output of a linear system is a very simple one. Once again a result in the frequency domain [equation (9.21)] is much simpler than the corresponding result in the time domain [equation (9.22)].

Theorem 9.2 can be used to evaluate the spectrum of some types of stationary process in a simpler way to that used in Chapter 6, where the method used was to evaluate the autocovariance function of the process and then find its Fourier transform. Several examples will now be given.

Moving average process   A MA process of order q is given by

$$X_t = \beta_0 Z_t + \ldots + \beta_q Z_{t-q}$$

where Z_t denotes a purely random process with variance σ_Z^2. This equation may be regarded as specifying a linear system with $\{Z_t\}$ as input and $\{X_t\}$ as output, whose frequency

206

response function is given by

$$H(\omega) = \sum_{j=0}^{q} \beta_j e^{-i\omega j}$$

As $\{Z_t\}$ is a purely random process, its spectrum is given by (see equation 6.19)

$$f_Z(\omega) = \sigma_Z^2/\pi$$

Thus, using (9.21), the spectrum of $\{X_t\}$ is given by

$$f_x(\omega) = \left| \sum_{j=0}^{q} \beta_j e^{-i\omega j} \right|^2 \sigma_Z^2/\pi$$

For example, for the first order MA process,

$$X_t = Z_t + \beta Z_{t-1}, \qquad (9.22a)$$

we have

$$H(\omega) = 1 + \beta e^{-i\omega}$$

and

$$G^2(\omega) = |H(\omega)|^2 = (1 + \beta \cos \omega)^2 + \beta^2 \sin^2 \omega$$
$$= 1 + 2\beta \cos \omega + \beta^2$$

so that $f_x(\omega) = (1 + 2\beta \cos \omega + \beta^2)\sigma_Z^2/\pi$ as already derived in Section 6.5.

This type of approach can also be used when $\{Z_t\}$ is not a purely random process. For example, suppose that the $\{Z_t\}$ process in equation (9.22a) has arbitrary spectrum $f_Z(\omega)$. Then the spectrum of $\{X_t\}$ is given by

$$f_x(\omega) = (1 + 2\beta \cos \omega + \beta^2)f_Z(\omega).$$

Autoregressive process   The first order AR process

$$X_t = \alpha X_{t-1} + Z_t$$

may be regarded as a linear system producing output X_t from input Z_t. It may also be regarded as a linear system 'the other

207

way round', producing output Z_t from input X_t by

$$Z_t = X_t - \alpha X_{t-1}$$

and this formulation which has frequency response function

$$H(\omega) = 1 - \alpha e^{-i\omega}$$

gives the desired result in the easiest mathematical way. Thus

$$G^2(\omega) = 1 - 2\alpha \cos \omega + \alpha^2$$

and so

$$f_Z(\omega) = (1 - 2\alpha \cos \omega + \alpha^2) f_x(\omega). \tag{9.23}$$

But if $\{Z_t\}$ denotes a purely random process with spectrum $f_Z(\omega) = \sigma_Z^2/\pi$, then equation (9.23) may be rewritten to evaluate $f_x(\omega)$ as

$$f_x(\omega) = \sigma_Z^2/\pi(1 - 2\alpha \cos \omega + \alpha^2),$$

which has already been obtained as equation (6.23) by the earlier method.

This approach may also be used for higher order AR processes.

Differentiation   Consider the linear system which converts a continuous input $X(t)$ into output $Y(t)$ by

$$Y(t) = \frac{\mathrm{d}X(t)}{\mathrm{d}t} \tag{9.24}$$

A differentiator is of considerable mathematical interest, although in practice only approximations to it are physically realizable.

If the input is sinusoidal, $X(t) = e^{i\omega t}$, then the output is given by

$$Y(t) = i\omega e^{i\omega t}$$

so that, using (9.18), the frequency response function is

$$H(\omega) = i\omega.$$

If the input is a stationary process, with spectrum $f_x(\omega)$, then it appears that the output has spectrum

$$f_y(\omega) = |i\omega|^2 f_x(\omega)$$
$$= \omega^2 f_x(\omega). \tag{9.25}$$

However this result assumes that the linear system (9.24) is stable, when in fact it is only stable for certain types of input process. For example, it is clear that the response to a unit step change is an unbounded impulse. In order for the system to be stable, the variance of the output must be finite. Now

$$\text{Var}[\,Y(t)] = \int_0^\infty f_y(\omega)\,\mathrm{d}\omega$$

$$= \int_0^\infty \omega^2 f_x(\omega)\,\mathrm{d}\omega.$$

But, using equation (6.18), we have

$$\gamma_x(k) = \int_0^\infty f_x(\omega)\cos\omega k\,\mathrm{d}\omega$$

and

$$\frac{\mathrm{d}^2\gamma_x(k)}{\mathrm{d}k^2} = -\int_0^\infty \omega^2 f_x(\omega)\cos\omega k\,\mathrm{d}\omega$$

so that

$$\text{Var}[\,Y(t)] = -\left[\frac{\mathrm{d}^2\gamma_x(k)}{\mathrm{d}k^2}\right]_{k=0}$$

Thus $Y(t)$ has finite variance provided that $\gamma_x(k)$ can be differentiated twice at $k = 0$, and only then does equation (9.25) hold.

9.3.5 Linear systems in series The advantages of working in the frequency domain are also evident when we consider two or more linear systems in series (or in cascade). For example Fig. 9.4 shows two linear systems in series, where the input $x(t)$ to system I produces output $y(t)$ which in turn is the input to system II producing output $z(t)$. It is often of interest to evaluate the properties of the overall system, which is also linear, where $x(t)$ is the input and $z(t)$ is the output. We will denote the impulse response and frequency response functions of systems I and II by $h_1(u)$, $h_2(u)$, $H_1(\omega)$, and $H_2(\omega)$.

Figure 9.4   Two linear systems in series.

In the time domain, the relationship between $x(t)$ and $z(t)$ would be in the form of a double integral involving $h_1(u)$ and $h_2(u)$, which is rather complicated. But in the frequency domain we can denote the Fourier transforms of $x(t)$, $y(t)$, $z(t)$ by $X(\omega)$, $Y(\omega)$, $Z(\omega)$ and use equation (9.20). Then

$$Y(\omega) = H_1(\omega)X(\omega)$$

and

$$Z(\omega) = H_2(\omega)Y(\omega)$$
$$= H_2(\omega)H_1(\omega)X(\omega).$$

Thus it is easy to see that the overall frequency response function of the combined system is

$$H(\omega) = H_1(\omega)H_2(\omega). \tag{9.26}$$

If

$$H_1(\omega) = G_1(\omega)e^{i\phi_1(\omega)}$$

210

and

$$H_2(\omega) = G_2(\omega)e^{i\phi_2(\omega)}$$

then

$$H(\omega) = G_1(\omega)G_2(\omega)e^{i[\phi_1(\omega) + \phi_2(\omega)]}$$

Thus the overall gain is the *product* of the component gains, while the overall phase is the *sum* of the component phases.

The above results are easily extended to the situation where there are k linear systems in series with respective frequency response functions, $H_1(\omega), \ldots, H_k(\omega)$. The overall frequency response function is

$$H(\omega) = H_1(\omega)H_2(\omega) \ldots \times H_k(\omega).$$

9.3.6 Design of filters The results of this section enable us to consider in more depth the properties of the filters introduced in Section 2.5.2. Given a time series, $\{x_t\}$, the filters for estimating or removing trend are of the form

$$y_t = \sum_k h_k x_{t-k}$$

and are clearly linear systems with frequency response function

$$H(\omega) = \sum_k h_k e^{-i\omega k}.$$

If the time-series has spectrum $f_x(\omega)$, then the spectrum of the smoothed series is given by

$$f_y(\omega) = G^2(\omega)f_x(\omega) \tag{9.27}$$

where $G(\omega) = |H(\omega)|$.

How do we set about choosing an appropriate filter for a time-series? The design of a filter involves a choice of $\{h_k\}$ and hence of $H(\omega)$ and $G(\omega)$. Two types of 'ideal' filter are shown in Fig. 9.5. Both have sharp cut-offs, the low-pass filter completely eliminating high frequency variation and the

Figure 9.5   Two types of ideal filter; (a) a low-pass filter or trend estimator; (b) a high-pass filter or trend eliminator.

high-pass filter completely eliminating low frequency variation.

But 'ideal' filters of this type are impossible to achieve with a finite set of weights. Instead the smaller the number of weights used, the less sharp will generally be the cut-off property of the filter. For example the gain diagram of a simple moving average of three successive observations (Fig. 9.2(c)) is of low-pass type but has a much less sharp cut-off than the 'ideal' low-pass filter (Fig. 9.5(a)). More sophisticated moving averages such as Spencer's 15 point moving average have much better cut-off properties.

The differencing filter (Section 2.5.3), for removing a trend, of the form

$$y_t = x_t - x_{t-1}$$

has frequency response function

$$H(\omega) = 1 - e^{-i\omega}$$

and gain function

$$G(\omega) = \sqrt{[2(1 - \cos \omega)]}$$

which is plotted in Fig. 9.6. This is indeed of high-pass type, but the cut-off property is rather poor and this should be borne in mind when working with first differences.

212

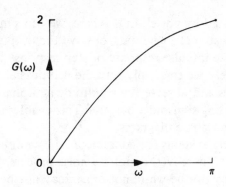

*Figure 9.6   The gain diagram for the
difference operator.*

9.4 The identification of linear systems

We have so far assumed that the structure of the linear system under consideration is known. Given the impulse response function of a system, or equivalently the frequency response function, we can find the output corresponding to a given input. In particular, when considering the properties of filters for estimating trend and seasonality, a formula for the 'system' is given.

But many problems concerning linear systems are of a completely different type. The structure of the system is *not* known and the problem is to examine the relationship between input and output so as to infer the properties of the system. This procedure is called the *identification* of the system. For example suppose we are interested in the effect of temperature on the yield from a chemical process. Here we have a physical system which we assume, initially at least, is approximately linear over the range of interest. By examining the relationship between observations on temperature (the input) and yield (the output) we can infer the properties of the chemical process.

The identification process is straightforward if the input to the system can be controlled and if the system is 'not

213

contaminated by noise'. In this case, we can simply apply an impulse or step change input, observe the output, and hence estimate the impulse response or step response function. Alternatively we can apply sinusoidal inputs at different frequencies and observe the amplitude and phase shift of the corresponding sinusoidal outputs. This enables us to evaluate the gain and phase diagrams.

But many systems are contaminated by noise as illustrated in Fig. 9.7, where $N(t)$ denotes a noise process. This noise process may not be white noise (i.e. may not be a purely random process), but is usually assumed to be uncorrelated with the input process, $X(t)$.

Figure 9.7   A linear system with added noise.

A further difficulty arises when the input is observable but not controllable. In other words one cannot make changes, such as a step change, to the input. For example attempts have been made to treat the economy as a linear system and to examine the relationship between variables like price increases (input) and wage increases (output). But price increases can only be controlled to a certain extent by governmental decisions, and there is also a feedback problem in that wage increases may in turn affect price increases (see Section 9.4.3).

When the system is affected by noise and/or the input is not controllable, more refined techniques are required to identify the system. We will describe two alternative approaches, one in the frequency domain and one in the time

domain. In Section 9.4.1, we show how cross-spectral
analysis of input and output may be used to estimate the
frequency response function of a linear system. In
Section 9.4.2, we describe a method proposed by Box and
Jenkins (1970) for estimating the impulse response function
of a linear system.

9.4.1 Estimating the frequency response function

Suppose that we have a linear system with added noise, as
depicted in Fig. 9.7, where the noise is assumed to be
uncorrelated with the input and to have mean zero. Suppose
also that we have observations on input and output over
some time-period and wish to estimate the frequency
response function of the system. We will denote the
(unknown) impulse response and frequency response
functions of the system by $h(u)$, $H(\omega)$ respectively.

The reader may think that equation (9.21), namely

$$f_y(\omega) = G^2(\omega)f_x(\omega)$$

can be used to estimate the gain of the system. But this
equation does not hold in the presence of noise, $N(t)$, and
does not in any case give information about the phase of the
system. Instead we will derive a relationship involving the
cross-spectrum of input and output.

In continuous time, the output $Y(t)$ is given by

$$Y(t) = \int_0^\infty h(u)X(t-u)du + N(t) \tag{9.28}$$

Note that we are only considering physically realizable
systems so that $h(u)$ is zero for $u < 0$. For mathematical
convenience we assume $E[X(t)] = 0$ so that $E[Y(t)] = 0$,
but the following results also hold if $E[X(t)] \neq 0$. Multiplying
through equation (9.28) by $X(t - \tau)$ and taking expectations,
we have

$$E[N(t)X(t-\tau)] = 0$$

215

since $N(t)$ is assumed to be uncorrelated with input, so that

$$\gamma_{xy}(\tau) = \int_0^\infty h(u) \, \gamma_{xx}(\tau - u) du \qquad (9.29)$$

where γ_{xy} is the cross-covariance function of $X(t)$ and $Y(t)$ and γ_{xx} is the autocovariance function of $X(t)$. Equation (9.29) is called the Wiener–Hopf integral equation and, given γ_{xy} and γ_{xx}, can in principle be solved to find the impulse response function, $h(u)$. But it is often easier to work with the corresponding relationship in the frequency domain.

First we revert to discrete time and note that the discrete-time analogue of equation (9.29) is

$$\gamma_{xy}(\tau) = \sum_{k=0}^\infty h_k \gamma_{xx}(\tau - k) \qquad (9.30)$$

Take Fourier transforms of both sides of this equation by multiplying by $e^{-i\omega\tau}/\pi$ and summing from $\tau = -\infty$ to $+\infty$. Then we find

$$f_{xy}(\omega) = \sum_{\tau=-\infty}^\infty \sum_{k=0}^\infty h_k e^{-i\omega k} \, \gamma_{xx}(\tau - k) e^{-i\omega(\tau-k)}/\pi$$

$$= \sum_{k=0}^\infty h_k e^{-i\omega k} f_x(\omega)$$

$$= H(\omega) f_x(\omega) \qquad (9.31)$$

where f_{xy} is the cross-spectrum of input and output and f_x is the (auto)spectrum of the input. Thus, once again, a convolution in the time domain corresponds to a multiplication in the frequency domain.

Estimates of $f_{xy}(\omega)$ and $f_x(\omega)$ can now be used to estimate $H(\omega)$ using (9.31). Denote the estimated spectrum of the input by $\hat{f}_x(\omega)$, and the estimate of f_{xy}, obtained by cross-spectral analysis, by $\hat{f}_{xy}(\omega)$. Then

$$\hat{H}(\omega) = \hat{f}_{xy}(\omega)/\hat{f}_x(\omega).$$

216

We usually write

$$H(\omega) = G(\omega)e^{i\phi(\omega)}$$

and estimate the gain and phase separately. We have

$$
\begin{aligned}
\hat{G}(\omega) &= |\hat{H}(\omega)| = |\hat{f}_{xy}(\omega)/\hat{f}_x(\omega)| \\
&= |\hat{f}_{xy}(\omega)|/\hat{f}_x(\omega) \qquad \text{since } f_x(\omega) \text{ is real} \\
&= \hat{\alpha}_{xy}(\omega)/\hat{f}_x(\omega)
\end{aligned}
\tag{9.32}
$$

where $\alpha_{xy}(\omega)$ is the cross-amplitude spectrum [see equation (8.13)].

We also find

$$\tan \hat{\phi}(\omega) = - \hat{q}(\omega)/\hat{c}(\omega) \tag{9.33}$$

where $q(\omega)$, $c(\omega)$ are the quadrature and co-spectra, respectively [see equation (8.14)].

Thus, having estimated the cross-spectrum, equations (9.32) and (9.33) enable us to estimate the gain and phase of the linear system, whether or not there is added noise.

We can also use cross-spectral analysis to estimate the properties of the noise process. The discrete-time version of equation (9.28) is

$$Y_t = \sum_{k=0}^{\infty} h_k X_{t-k} + N_t \tag{9.34}$$

For mathematical convenience, we again assume that

$$E(N_t) = E(X_t) = 0 \text{ so that } E(Y_t) = 0.$$

If we multiply both sides of (9.34) by Y_{t-m}, we find

$$Y_t Y_{t-m} = (\Sigma h_k X_{t-k} + N_t)(\Sigma h_k X_{t-m-k} + N_{t-m})$$

Taking expectations we find

$$\gamma_{yy}(m) = \sum_k \sum_j h_k h_j \gamma_{xx}(m - k + j) + \gamma_{nn}(m)$$

since (X_t) and (N_t) are assumed to be uncorrelated.

217

Taking Fourier transforms of both sides of this equation, we find

$$f_y(\omega) = H(\omega)\,\overline{H(\omega)}\,f_x(\omega) + f_n(\omega)$$

But $H(\omega)\,\overline{H(\omega)} = G^2(\omega)$

$$= C(\omega)f_y(\omega)/f_x(\omega)$$

so that

$$f_n(\omega) = f_y(\omega)[1 - C(\omega)] \qquad (9.35)$$

Thus an estimate of $f_n(\omega)$ is given by

$$\hat{f}_n(\omega) = \hat{f}_y(\omega)[1 - \hat{C}(\omega)].$$

Equation (9.35) also enables us to see that if there is no noise, so that there is a pure linear relation between X_t and Y_t, then $f_n(\omega) = 0$ and $C(\omega) = 1$ for all ω. On the other hand if $C(\omega) = 0$ for all ω, then $f_y(\omega) = f_n(\omega)$ and the output is not linearly related to the input. This confirms the point mentioned in Chapter 8 that the coherency, $C(\omega)$, measures the linear correlation between input and output at frequency ω.

The results of this section, not only show us how to identify a linear system by cross-spectral analysis, but also give further guidance on the interpretation of functions derived from the cross-spectrum, particularly the gain, phase and coherency. An example, involving a chemical process, is given by Goodman *et al.* (1961), while Gudmundsson (1971) describes an economic application of cross-spectral analysis.

In principle, estimates of the frequency response function of a linear system may be transformed to give estimates of the impulse response function (Jenkins and Watts, 1968, p. 444) but I do not recommend this. For instance example 2 of Section 8.2.1 appears to indicate that the sign of the phase may be used to indicate which series is 'leading the other. But

Hause (1971) has shown that for more complicated lagged models of the form

$$Y_t = \sum_{k=0}^{\infty} h_k X_{t-k}, \tag{9.36}$$

which are called *distributed lag* models by economists, it becomes increasingly difficult to make inferences from phase estimates and Hause concludes that phase leads and lags will rarely provide economists with direct estimates of the *time* domain relationships that are of more interest to them.

9.4.2 The Box—Jenkins approach   This section gives a brief introduction to the method proposed by G. E. P. Box and G. M. Jenkins for identifying a physically realizable linear system, in the time domain, in the presence of added noise. Further details can be found in Box and Jenkins (1968, 1970).

The input and output series are both differenced d times until both are stationary, and are also mean-corrected. The modified series will be denoted by $\{X_t\}$, $\{Y_t\}$, respectively. We want to find the impulse response function, $\{h_k\}$, of the system where

$$Y_t = \sum_{k=0}^{\infty} h_k X_{t-k} + N_t \tag{9.37}$$

The 'obvious' way to estimate $\{h_k\}$ is to multiply through equation (9.37) by X_{t-m} and take expectations to give

$$\gamma_{xy}(m) = h_0 \gamma_{xx}(m) + h_1 \gamma_{xx}(m-1) + \ldots \tag{9.38}$$

assuming that N_t is uncorrelated with the input. If we assume that the weights $\{h_k\}$ are effectively zero beyond $k = K$, then the first $(K + 1)$ equations of the type (9.38) for $m = 0, 1, \ldots,$ K, can be solved for the $(K + 1)$ unknowns $h_0, h_1, \ldots, h_K$, on substituting estimates of γ_{xy} and γ_{xx}. Unfortunately these equations do not, in general, provide good estimators for the $\{h_k\}$, and, in any case, assume knowledge of the truncation point K. The basic trouble, as already noted in Section 8.1.2, is that autocorrelation within the input and output series will increase the variance of cross-correlation estimates.

Box and Jenkins (1970, Chapter 1) therefore propose two modifications to the above procedure. Firstly, they suggest 'prewhitening' the input before calculating the sample cross-covariance function. Secondly, they propose an alternative form of equation (9.36) which will in general require fewer parameters. They represent the linear system by the equation

$$Y_t - \delta_1 Y_{t-1} - \ldots - \delta_r Y_{t-r} = \omega_0 X_{t-b}$$
$$- \omega_1 X_{t-b-1} \ldots - \omega_s X_{t-b-s}$$
$$(9.39)$$

This is rather like equation (9.4), but is given in the notation used by Box and Jenkins (1970) and involves an extra parameter, b, which is called the *delay* of the system. The delay can be any non-negative integer. Using the backward shift operator, B, (9.39) may be written as

$$\delta(B)Y_t = \omega(B)X_{t-b} \qquad (9.40)$$

where

$$\delta(B) = 1 - \delta_1 B - \ldots - \delta_r B^r$$

and

$$\omega(B) = \omega_0 - \omega_1 B - \ldots - \omega_s B^s.$$

Box and Jenkins (1970) describe equation (9.40) as a *transfer function* model, which is not a good description in that many authors use the term transfer function to describe the frequency response function [equation (9.10)].

The Box–Jenkins procedure begins by fitting an ARMA model to the (differenced) input. Suppose this model is of the form (see Section 3.4.5)

$$\phi(B)X_t = \theta(B)\alpha_t$$

where $\{\alpha_t\}$ denotes a purely random process, in the notation of Box and Jenkins. Thus we can transform the input to a white noise process by

$$\phi(B)\theta^{-1}(B)X_t = \alpha_t.$$

Suppose we apply the same transformation to the output, to give

$$\phi(B)\theta^{-1}(B)Y_t = \beta_t$$

and then calculate the cross-covariance function of the filtered input and output, namely (α_t) and (β_t). It turns out that this function gives a better estimate of the impulse response function, since, if we write

$$h(B) = h_0 + h_1 B + h_2 B^2 + \ldots$$

so that

$$Y_t = h(B)X_t + N_t,$$

then

$$\begin{aligned}
\beta_t &= \phi(B)\theta^{-1}(B)Y_t \\
&= \phi(B)\theta^{-1}(B)[h(B)X_t + N_t] \\
&= h(B)\,\alpha_t + \phi(B)\theta^{-1}(B)N_t
\end{aligned}$$

and

$$\gamma_{\alpha\beta}(m) = h_m\ \text{Var}(\alpha_t) \tag{9.41}$$

since (α_t) is a purely random process, and N_t is uncorrelated with (α_t). Equation (9.41) is of a much simpler form to equation (9.38). If we denote the sample cross-covariance function of α_t and β_t by $c_{\alpha\beta}$, and the observed variance of α_t by s_α^2, then an estimate of h_m is given by

$$\hat{h}_m = c_{\alpha\beta}(m)/s_\alpha^2. \tag{9.42}$$

These estimates should be more reliable than those given by the solution of equations of type (9.38).

Box and Jenkins (1970) give the theoretical impulse response functions for a variety of models of type (9.39) and go on to show how the shape of the estimated impulse response function given by (9.42) can be used to suggest appropriate values of the integers r, b, and s in equation (9.39). They then show how to obtain least squares estimates of δ_1, δ_2, . . . , ω_0, ω_1, . . . , given values of r, b, and s. These estimates can in turn be used to obtain refined estimates of (h_m) if desired.

Box and Jenkins go on to show how a model of type (9.39), with added noise, can be used for forecasting and control. A few case studies have now been published (e.g. in Jenkins, 1979) and the method looks potentially useful, though it is too early to give a final assessment of the method. However, it should be noted that the main example discussed by Box and Jenkins (1970) using some gas furnace data has been criticized by Young (1974) and Chatfield (1977, p. 504) on a number of grounds including the exceptionally high correlation between input and output which means that virtually any identification procedure will give good results.

A modification to the Box—Jenkins procedure has been proposed by Haugh and Box (1977) in which separate models are fitted to both the input and output before cross-correlating the resulting residual series. Further experience is needed to see if this is generally advisable. The modified procedure has been used by Pierce (1977) to assess the relationship between various pairs of economic series, and he shows that, after taking account of autocorrelation within each series, there is usually very little correlation left between the residual series. Pierce discusses how to reconcile these results with the views of economists.

The question as to when it is better to use cross-spectral analysis or a time-domain parametric model approach is still unresolved though it has been discussed by a number of authors (e.g. Astrom and Eykhoff, 1971). It seems to me to

be rather dangerous to attempt to make general pronounce-
ments on the relative virtues of time-domain and frequency-
domain methods, though Box and Jenkins clearly favour a
time-domain approach. The two approaches appear to be
complementary rather than rivals, and it may be helpful to
try both.

In conclusion, it is worth noting that a similar method to
the Box—Jenkins approach has been independently
developed by two control engineers, K. J. Astrom and T.
Bohlin. This method also involves prewhitening and a model
of type (9.39) (Astrom and Bohlin, 1966; Astrom, 1970),
but does not discuss identification and estimation procedures
in equal depth. One difference of the Astrom—Bohlin
method is that non-stationary series may be converted to
stationarity by methods other than differencing. The results
of Clarke (1973) suggest that high-pass filtering or linear
trend removal may sometimes be better than differencing.

9.4.3 Systems involving feedback   A system of the type
illustrated in Fig. (9.7) is called an *open-loop* system, and the
procedures described in the previous two sections are
appropriate for data collected under these conditions. But
data is often collected from systems where some form of
feedback control is being applied, and then we have what is
called a *closed-loop* system as illustrated in Fig. 9.8. For
example, when trying to identify a full-scale industrial
process, it could be dangerous, or an unsatisfactory product
could be produced, unless some form of feedback control is
applied to keep the output somewhere near target. Similar
problems arise in an economic context. For example, attempts
to find a linear relationship showing the effect of price
changes on wage changes are bedevilled by the fact that wage
changes will in turn affect prices.

The problem of identifying systems in the presence of
feedback control is discussed by Granger and Hatanaka
(1964, Chapter 7), Astrom and Eykhoff (1971, p. 130),

Figure 9.8   A closed-loop system.

Priestley (1971), Box and Macgregor (1974), and Gustavsson *et al.* (1977), and it is important to realise that open-loop procedures may not be applicable. The situation can be explained more clearly in the frequency domain. Let $f_{xy}(\omega)$ denote the cross-spectrum of $X(t)$ and $Y(t)$ in Fig. 9.8, and let $f_x(\omega)$, $f_n(\omega)$, $f_v(\omega)$ denote the spectra of $X(t)$, $N(t)$, and $V(t)$ respectively. Then if $H(\omega)$ and $J(\omega)$ denote the frequency response functions of the system and controller respectively, it can be shown (e.g. Akaike, 1967) that

$$f_{xy}/f_x = (Hf_v + \bar{J}f_n)/(f_v + J\bar{J}f_n) \tag{9.43}$$

where all terms are functions of frequency, and $\bar{J}$ is the complex conjugate of J. Only if $f_n \equiv 0$ or $J \equiv 0$, is the ratio f_{xy}/f_x equal to H as is the case for an open-loop system (equation 9.31). Thus the estimate of H provided by $\hat{f}_{xy}/\hat{f}_x$ will be poor unless f_n/f_v is small (Priestley, 1971). In particular if $f_v \equiv 0$, $\hat{f}_{xy}/\hat{f}_x$ will provide an estimate of J^{-1} and *not* of H.

Similar remarks apply to an analysis in the time domain. The time-domain equivalent of (9.43) is given by Box and MacGregor (1974).

The above problem is not specifically discussed by Box and Jenkins (1970), although it is quite clear from the remarks in their Section 11.6 that their methods are only intended for use in open-loop systems. However some confusion may be created by the fact that Box and Jenkins (1970; Section 12.2) do discuss ways of choosing optimal feedback control which is in fact quite a different problem. Having identified a system in open-loop, they show how to choose the feedback control action so as to satisfy some chosen criterion.

Unfortunately open-loop identification procedures have sometimes been used for a closed-loop system where they are not appropriate. Tee and Wu (1972) studied a paper machine while it was already operating under manual control and proposed a control procedure which has been shown to be worse than the existing form of control (Box and MacGregor, 1974). In marketing, several authors have investigated the relationship between expenditure on offers and advertising on the sales of products such as washing-up liquid and coffee. However, expenditure on advertising is often in turn affected by changes in sales levels, so that any conclusions obtained by an open-loop analysis are open to doubt.

What then can be done if feedback is present? Box and MacGregor (1974) suggest one possible approach in which one deliberately adds an independent programmed noise sequence on top of the noise, $V(t)$. Alternatively one may have some knowledge of the noise structure or of the controller frequency response function. Akaike (1968) claims that it is possible to identify a system provided only that instantaneous transmission of information does not occur in both system and controller, and an example of his, rather complicated, procedure is given by Otomo et al. (1972).

However a further difficulty is that it is not always clear if feedback is present or not, particularly in economics and marketing. Some indication may be given by the methods of Granger and Hatanaka (1964, Chapter 7), or by observing significantly large cross-correlation coefficients between (prewhitened) input and output at a zero or positive lag.

Alternatively it may be clear from physical considerations that feedback is or is not present. For example in studying the relationship between average (ambient) temperature and sales of a product, it is clear that sales cannot possibly affect temperature so that one has an open-loop system. But if feedback is present, then standard cross-correlation and cross-spectral analysis should not be used as they may give spurious results.

Exercises

1. Find the impulse response function, the step response function, the frequency response function, the gain and the phase shift for the following linear systems (or filters).

   (a) $y_t = \frac{1}{2}x_{t-1} + x_t + \frac{1}{2}x_{t+1}$

   (b) $y_t = \frac{1}{5}(x_{t-2} + x_{t-1} + x_t + x_{t+1} + x_{t+2})$

   (c) $y_t = \nabla x_t$

   (d) $y_t = \nabla^2 x_t$

   where in each case t is integer-valued. Plot the gain and phase shift for filters (a) and (c). Which of the filters are (i) low-pass (ii) high-pass?

   If the filters (a) and (b), given above, are joined in series, find the frequency response function of the combined filter.

2. Find the frequency response functions of the following linear systems in continuous time.

   (a) $y(t) = gx(t - \tau)$

   where g, τ are positive constants

   (b) $y(t) = \frac{g}{T} \int_0^\infty e^{-u/T} x(t - u) du$

3.   If $\{X_t\}$ is a stationary discrete time-series with power spectral density function $f(\omega)$, show that the smoothed time-series

$$Y_t = \sum_{p=0}^{k} a_p X_{t-p},$$

where the a's are real constants, is a stationary process which has power spectral density function

$$\left[\sum_{q=0}^{k} \sum_{p=0}^{k} a_p a_q \cos(p-q)\omega \right] f(\omega).$$

In particular if $a_p = 1/(k+1)$ for $p = 0, 1, \ldots, k$, show that the power spectrum of Y_t is

$$f(\omega)[1 - \cos(k+1)\omega]/(k+1)^2(1 - \cos \omega).$$

[Hint: Use equation (9.21) and the trigonometric relation $\cos A \cos B = \frac{1}{2}[\cos(A+B) + \cos(A-B)]$.]

4.   Consider the one-parameter second-order AR process

$$X_t = \alpha X_{t-2} + Z_t$$

where $\{Z_t\}$ denotes a purely random process, mean zero, variance σ_Z^2. Show that the process is second-order stationary if $|\alpha| < 1$, and find its autocovariance and autocorrelation functions.

Show that the power spectral density function of the process is given by

$$f(\omega) = \sigma_Z^2/\pi(1 - 2\alpha \cos 2\omega + \alpha^2) \quad (0 < \omega < \pi)$$

using two different methods: a) By transforming the autocovariance function. b) By using the approach of Section 9.3.4.

Suppose now that $\{Z_t\}$ is any stationary process which has power spectrum $f_Z(\omega)$. What then is the power spectrum of $\{X_t\}$ as defined by the above equation?

227

CHAPTER 10

Some other Topics

This chapter provides a brief introduction and references to a number of topics, including some recent developments, which are not covered in the earlier part of this book. Other recent advances are reviewed by Chatfield (1977), Cox (1981), Newbold (1981) and the collection of papers edited by Brillinger and Tiao (1980).

CONTROL THEORY   To many statisticians, the word 'control' implies statistical quality control using control charts, but in this section we use the word as in control engineering to denote the search for an automatic control procedure for a system whose structure may or may not be known. This is often the ultimate objective in identifying linear systems as in Chapter 9.

There are many different approaches to control such as those based on linear parametric models of the form (c.f. (9.40))

$$\delta(B)Y_t = \omega(B)X_t + \theta(B)Z_t$$

where $\{Z_t\}$ denotes a purely random process and δ, ω, θ are polynomials in the backward shift operator, B. Many other approaches have been described including the use of cross-spectral analysis, and stochastic adaptive systems (e.g. Harris, 1976, Section 7) but we do not have space to discuss them all here. The main contributions to control theory have naturally been made by control engineers, though mathematicians, statisticians and operational researchers have also played a part.

Control engineers were originally concerned mainly with

deterministic control (and many of them still are). Some of the reported research, such as the solution of non-linear differential equations subject to boundary conditions, has an obvious relevance to control theory but may equally be regarded as a branch of applied mathematics. In recent years attention has moved away from deterministic problems to the consideration of stochastic control where the system being controlled is subject to random disturbances. Some references covering the whole field include Solodovnikov (1965), Astrom (1970), Fuller (1970), Jacobs (1974), Harris (1976) and Priestley (1981).

In stochastic control, a basic problem is that of separating the signal from the noise (see Section 5.6) and there has been much work on the filtering problem starting with the work of Wiener and Kolmogorov. One major development has been a filter, called the Kalman filter, which is a recursive method of estimating the state of a system in the presence of noise. The reader is referred to Meinhold and Singpurwalla (1983), Young (1974) and Priestley (1981, Section 10.6). Kalman filtering has been used in many applications including the control of a space rocket where the system dynamics are well defined but the disturbances are unknown.

Although many statistical problems arise in control theory, statisticians have made a relatively small contribution to the subject. Box and Jenkins (1970) are one exception and these authors show how to identify a linear parametric model for a system and hence find a 'good' control procedure. It is interesting to note that a somewhat similar approach has been developed by two control engineers in Astrom and Bohlin (1966), though the latter do not give such detailed guidance on estimation, diagnostic-checking and the treatment of non-stationary series. The Box—Jenkins approach has been compared with the control engineer's state space representation approach by Priestley (1981). There has been regrettably little communication between statisticians and control engineers. Of course, there are differences in emphasis and interest.

229

Statisticians tend to work in discrete time, while control engineers often prefer continuous time. Box and Jenkins (197? stress the need to identify systems while much of the control literature appears to assume knowledge of the system structur? Nevertheless it is to be hoped that the future will bring more collaboration.

NON-STATIONARY SERIES   Much of the available theory is applicable to *stationary* processes, although most real time series do change with time, albeit slowly in some cases. We have already discussed some methods of transforming data t? stationarity, such as differencing, in order to use this theory and fit stationary models. However, it is often more important to describe the non-stationary features of the series, such as trend and seasonality, rather than the properties of the stationary residuals.

With a long, apparently stationary series, it is still a good idea to split the series into two or more segments and compar? the properties of each segment, particularly the form of the ac.f. and spectrum. One approach to slowly-changing series is to fit a time-dependent spectrum called an *evolutionary spectrum*, where the frequency properties are examined for *overlapping* segments. This topic is reviewed by Priestley (1981, Chapter 11) who also introduces some other definitions of time-dependent spectra. Cox (1981, Section 4) also reviews non-stationary processes.

A different kind of non-stationarity arises when there is a sudden change in a time series due to the occurrence of a known external event such as a wage freeze. Box and Tiao (1975) model such changes with a technique called *intervention analysis* which is similar to the use of dummy variables in regression.

NON-LINEAR MODELS   Much of the time-series literature not only assumes that all series are stationary (or can be

230

SOME OTHER TOPICS

educed to stationarity by a simple transformation such as
differencing) but also assumes that the appropriate model
will be *linear*, so that the observed series can be represented
as a linear function of the present and past values of a purely
random process. For example the AR, MA, ARMA and
ARIMA models are all linear.

Recently certain types of non-linear model have been
investigated (Granger, 1978; Priestley, 1981) and these seem
likely to be further developed in the next few years. One
class of non-linear models is called the *bilinear* class which
may be regarded as the natural non-linear extension of an
ARMA model. For example the first-order bilinear model is
given by

$$X_t = aX_{t-1} + bZ_t + cZ_{t-1}X_{t-1}$$

where $\{Z_t\}$ denotes a purely random process and a, b, c are
parameters. It is the last term on the right-hand side of the
above equation which is the non-linear term.

Another class of models are threshold autoregressive (TAR)
models where the model parameters depend on the past
values of the process. For example, a 1st-order TAR model
would be

$$X_t = \begin{cases} \alpha_1 X_{t-1} + Z^{(1)}_{t-1} & \text{if } X_{t-1} < d \\[2ex] \alpha_2 X_{t-1} + Z^{(2)}_{t-1} & \text{if } X_{t-1} \geq d \end{cases}$$

MODEL IDENTIFICATION TOOLS   The two standard
tools which are used to identify an appropriate ARMA model
for a given stationary time-series are the sample ac.f. and the
sample partial ac.f. (see Chapter 4). Recently several new
tools have been proposed for this purpose and in particular
for selecting the appropriate order of an AR process (see
Section 4.2.2).

An alternative to the partial ac.f. is the inverse ac.f. whose

231

use in identifying ARMA models is described by Chatfield (1979). The inverse ac.f. of the ARMA model (see Equation 3.5a)

$$\phi(B)X_t = \theta(B)Z_t$$

is exactly the same as the ordinary ac.f. of the corresponding inverse ARMA model given by

$$\theta(B)X_t = \phi(B)Z_t$$

where ϕ and θ are interchanged. It turns out that the inverse ac.f. has similar properties to the partial ac.f. in that it 'cuts off' at lag p for an AR(p) process but generally dies out slowly for MA and ARMA processes. In other words it has inverse properties to the ordinary ac.f. which 'cuts off' for MA processes. Chatfield (1979) argues that the inverse ac.f. often contains more information than the partial ac.f., extends easily to seasonal ARIMA models, and may well displace the partial ac.f.

Instead of subjectively examining functions like the ac.f., an alternative type of approach is to choose the ARMA model which optimizes a suitably chosen function of the data. One approach, based on Akaike's final prediction error (FPE) criterion, is concerned with comparing AR processes of different order. The order, p, is essentially selected so as to minimize an unbiased form of the residual mean square when multiplied by the quantity $\{1 + (1 + p)/N\}$ which is chosen on theoretical grounds so as to ensure that the minimum occurs at about the right value (Jones, 1974, 1975). A more general criterion for model selection is to minimize a quantity called Akaike's information criterion (AIC), which is equal to [-2 ln (maximized likelihood) + 2 (number of independent parameters estimated)]. The AIC criterion can be used to compare ARMA models as well as AR models. A third criterion is Parzen's autoregressive transfer function criterion (CAT) which also tends to give similar results (e.g. Landers and Lacoss, 1977). Unfortunately

all these criteria may give more than one minimum, depend
on assuming that the data are normally distributed, and
sometimes indicate too many parameters (Shibata, 1976).
Thus they should be used only as guides.

AUTOREGRESSIVE SPECTRUM ESTIMATION Spectral
analysis is concerned with estimating the spectrum of a
stationary stochastic process. The approach described in
Chapter 7, which is based on Fourier analysis, is essentially
non-parametric in that no model is assumed *a priori*. In
recent years a parametric approach, called autoregressive
spectrum estimation has become a popular alternative
method and this will now be briefly described.

Many stationary stochastic processes can be adequately
approximated by an AR process of sufficiently high order,
say

$$X_t = \alpha_0 + \alpha_1 X_{t-1} + \ldots + \alpha_p X_{t-p} + Z_t \qquad (10.1)$$

The spectrum of (10.1) at frequency ω is inversely
proportional to

$$| 1 - \sum_{k=1}^{p} \alpha_k e^{-ik\omega} |^2.$$

The procedure requires the user to assess the order, p, of the
process, possibly using Akaike's information criterion. Then
the AR parameters are estimated, as described in Section 4.2,
and substituted into the expression for the spectrum of
(10.1). The approach tends to give a smoother spectrum than
that given by the non-parametric approach, though it can also
pick out fairly narrow peaks in the spectrum. Convincing
examples are given for example by Jones (1974), Griffiths
and Prieto-Diaz (1977), and Tong (1977), and the approach is
certainly worth considering. Note that the selected order, p,
of the process may be higher than might be expected as
values around 10 are common and values as high as 28 have
been reported.

233

An obvious development of this approach is to fit ARMA models rather than AR models to give what might be called ARMA spectrum estimation, but the simplicity of fitting AR models is likely to inhibit such a development.

SPATIAL SERIES   In some scientific areas, particularly in ecology and agriculture, data often arise which are ordered with respect to one or two *spatial* co-ordinates rather than with respect to time. Methods of analysing spatial data have been described, for example, by Besag (1974) and Cliff and Ord (1975). Although there are many similarities between time-series and spatial analysis, there are also fundamental differences (Chatfield, 1977, Section 9). Ripley (1981) provides an excellent review of spatial statistics.

'CROSSING' PROBLEMS   If one draws a horizontal line (an axis) through a stationary time series, then the time series will cross and re-cross the line in alternate directions. The times at which the series cuts the axis provide valuable information about the properties of the series. Indeed in some practical problems the crossing points are the only information available. The problems of inferring the properties of a time series from its 'level-crossing' properties are discussed by Cramer and Leadbetter (1967), and a comprehensive survey and bibliography is given by Blake and Lindsey (1973).

THE CEPSTRUM   This word, which is an anagram of spectrum, is the power spectrum of the logarithmic power spectrum of a given time series. If a time series contains an echo, with a constant delay time, then the analysis of a finite record will result in a periodic spectrum with peaks at the harmonics which are integer multiples of the fundamental frequency. A spectral analysis of such a spectrum may be a helpful step in the analysis. The term 'cepstrum' was invented by J. W. Tukey and introduced in a paper by Bogert *et al.* (1963). This paper also coined such terms as 'quefrency' and 'saphe' which have 'obvious' meanings. Another reference is Noll (1973).

OBSERVATIONS AT UNEQUAL INTERVALS   This book has been mainly concerned with discrete time series measured at equal intervals of time. When observations are taken at unequal intervals, either by accident or design, the general definition of the periodogram (equation 7.17) still applies but the values of t are no longer integers. Autocorrelation coefficients can only be obtained directly by fitting a smoothed approximation to the time series. *Splines*, which are piecewise polynomials, can be used to provide such an approximation (see Wold, 1974). Further references on time series with unequally spaced or missing observations are given by Priestley (1981, p. 586) and Hannan (1970, p. 48); see also Roberts and Gaster (1980).

POLYSPECTRA   Brillinger (1965, 1975) discusses the higher-order spectra, or *polyspectra*, of multivariate stationary time series which are the Fourier transform of the appropriate cumulant. For a single time series, the first-order polyspectrum is the usual power spectrum, while the second-order polyspectrum is the *bispectrum*, discussed by Godfrey (1965), which may be able to determine the form of a non-linear element. For a pair of time series, the first-order polyspectrum is the usual cross-spectrum. I have no practical experience with polyspectra and will not pursue the topic.

MULTIVARIATE SERIES   Observations are often taken simultaneously on more than one time series. For example, in meteorology we might observe temperature, air pressure, rainfall, etc., at the same site over the same period of time. In examining the interrelationships among the different series, it is useful to distinguish between the case where the variables arise on an equal footing and the case where one wants to 'explain' variation in one series in terms of variation in other series. The presence of autocorrelation within each series and the possible presence of feedback between series makes the analysis difficult. The reader is referred to Priestley (1981, Volume 2) and Tiao and Box (1981). The case of two time series has already been considered in Chapters 8 and 9 though even here the author must admit that he has typically found it difficult to interpret cross-correlation and cross-spectral estimates. The analysis of three or more series is in theory a natural extension but in practice much more difficult and not recommended for the beginner.

The Fourier, Laplace and Z-transforms

This appendix provides a brief introduction to the Fourier transform which is a valuable mathematical tool in time-series analysis. Further details may be found in many books such as Hsu (1967) and Sneddon (1972). The related Laplace and Z-transforms are also introduced.

Given a (possibly complex-valued) function $h(t)$ of a real variable t, the Fourier transform of $h(t)$ is usually defined as

$$H(\omega) = \int_{-\infty}^{\infty} h(t)e^{-i\omega t}dt \tag{A1.1}$$

provided the integral exists for every real ω. Note that $H(\omega)$ is in general complex. A sufficient condition for $H(\omega)$ to exist is

$$\int_{-\infty}^{\infty} |h(t)| \, dt < \infty.$$

If (A1.1) is regarded as an integral equation for $h(t)$ given $H(\omega)$, then a simple inversion formula exists of the form

$$h(t) = \frac{1}{2\pi} \int_{-\infty}^{\infty} H(\omega)e^{i\omega t}d\omega \tag{A1.2}$$

and $h(t)$ is called the inverse Fourier transform of $H(\omega)$, or sometimes just the Fourier transform of $H(\omega)$. The two functions $h(t)$ and $H(\omega)$ are commonly called a Fourier transform pair.

The reader is warned that many authors use a slightly different definition of a Fourier transform to A1.1. For example some authors put a constant $1/\sqrt{2\pi}$ outside the integral in (A1.1) and then the inversion formula for $h(t)$ is symmetric. In time-series analysis many authors (e.g. Cox and Miller, 1968, p. 315) put a constant $1/2\pi$ outside the integral in A1.1. The inversion formula then has a unity constant outside the integral.

Some authors in time-series analysis (e.g. Jenkins and Watts, 1968, Blackman and Tukey, 1959) define Fourier transforms in terms of the variable $f = \omega/(2\pi)$ rather than ω. We then find that the Fourier transform pair is

$$G(f) = \int_{-\infty}^{\infty} h(t)e^{-2\pi i f t} dt \tag{A1.3}$$

$$h(t) = \int_{-\infty}^{\infty} G(f)e^{2\pi i f t} df \tag{A1.4}$$

Note that the constant outside each integral is now unity.

In time-series analysis, we will often use the discrete form of the Fourier transform when $h(t)$ is only defined for integer values of t. Then

$$H(\omega) = \sum_{t=-\infty}^{\infty} h(t)e^{-i\omega t} \qquad (-\pi \leqslant \omega \leqslant \pi) \tag{A1.5}$$

is the discrete Fourier transform of $h(t)$. Note that $H(\omega)$ is only defined in the interval $[-\pi, \pi]$. The inverse transform is

$$h(t) = \frac{1}{2\pi} \int_{-\pi}^{\pi} H(\omega)e^{i\omega t} d\omega \tag{A1.6}$$

Fourier transforms have many useful properties, some of which are used during the later chapters of this book.

However we will not attempt to review them here. The reader is referred for example to Hsu (1967).

One special type of Fourier transform arises when $h(t)$ is a real-valued even function such that $h(t) = h(-t)$. The autocorrelation function of a stationary time series has these properties. Then using (A1.1) with a constant $1/\pi$ outside the integral, we find

$$H(\omega) = \frac{1}{\pi} \int_{-\infty}^{\infty} h(t)e^{-i\omega t} dt$$

$$= \frac{2}{\pi} \int_{0}^{\infty} h(t) \cos \omega t \, dt \qquad (A1.7)$$

and it is clear that $H(\omega)$ is a real–valued even function. The inversion formula is then

$$h(t) = \frac{1}{2} \int_{-\infty}^{\infty} H(\omega)e^{i\omega t} d\omega$$

$$= \int_{0}^{\infty} H(\omega)\cos \omega t \, d\omega \qquad (A1.8)$$

Equations (A1.7) and (A1.8) are similar to a Fourier transform pair and are useful when we only wish to define $H(\omega)$ for $\omega > 0$. This pair of equations appear as equations (2.73) and (2.74) in Yaglom (1962). When $h(t)$ is only defined for integer values of t, equations (A1.7) and (A1.8) become

$$H(\omega) = \frac{1}{\pi} \left\{ h(0) + 2 \sum_{t=1}^{\infty} h(t) \cos \omega t \right\} \qquad (A1.9)$$

$$h(t) = \int_{0}^{\pi} H(\omega) \cos \omega t \, d\omega \qquad (A1.10)$$

and $H(\omega)$ is now only defined on $[0, \pi]$.

The Laplace transform of a function $h(t)$ which is defined

for $t > 0$ is given by

$$H(s) = \int_0^\infty h(t)e^{-st}\,dt \tag{A1.11}$$

where s is a complex variable. The integral converges when the real part of s exceeds some number called the abscissa of convergence.

The relationship between the Fourier and Laplace transforms is of some interest, particularly as control engineers often prefer to use the Laplace transform when investigating the properties of a linear system (e.g. Solodovnikov, 1965) as this will cope with physically realizable systems which are stable *or* unstable. If the function $h(t)$ is such that

$$h(t) = 0 \qquad (t < 0) \tag{A1.12}$$

and the real part of s is zero, then the Laplace and Fourier transforms of $h(t)$ are the same. The impulse response function of a physically realizable linear system satisfies (A1.12) and so for such functions the Fourier transform is a special case of the Laplace transform. More details about the Laplace transform may be found in many books, (e.g. Sneddon, 1972).

The Z-transform of a function $h(t)$ defined on the non-negative integers is given by

$$H(z) = \sum_{t=0}^\infty h(t)z^{-t} \tag{A1.13}$$

In discrete time, with a function satisfying (A1.12), some authors prefer to use the Z-transform rather than the discrete form of the Fourier transform (i.e. A1.5) or the discrete form of the Laplace transform, namely

$$H(s) = \sum_{t=0}^\infty h(t)e^{-st} \tag{A1.14}$$

Comparing (A1.13) with (A1.14) we see that $z = e^s$. The reader will observe that when $\{h(t)\}$ is a probability function such that $h(t)$ is the probability of observing the value t, for $t = 0, 1, \ldots$, then (A1.13) is related to the probability generating function of the distribution, while (A1.5) and (A1.14) are related to the moment generating function.

Exercises

1. If $h(t)$ is real, show that the real and imaginary parts of its Fourier transform, as defined by equation (A1.1), are even and odd functions respectively.

2. If $h(t) = e^{-a|t|}$ for all real t, where a is a positive real constant, show that its Fourier transform, as defined by equation (A1.1), is given by

$$H(\omega) = 2a/(a^2 + \omega^2) \qquad (-\infty < \omega < \infty)$$

3. Show that the Laplace transform of $h(t) = e^{-at}$ $(t > 0)$, where a is a real constant, is given by

$$H(s) = 1/(s + a) \qquad \text{Re}(s) > a.$$

APPENDIX II

The Dirac Delta Function

Suppose that $\phi(t)$ is any function which is continuous at $t = 0$. Then the *Dirac delta function,* $\delta(t)$, is such that

$$\int_{-\infty}^{\infty} \delta(t)\phi(t)\mathrm{d}t = \phi(0) \tag{A2.1}$$

Because it is defined in terms of its integral properties alone, it is sometimes called the 'spotting' function since it picks out one particular value of $\phi(t)$. It is also sometimes called simply the *delta function.*

It is important to realize that $\delta(t)$ is *not* a function. Rather it is a generalized function, or distribution, which maps a function into the real line.

Some authors define the delta function by

$$\delta(t) = \begin{cases} 0 & t \neq 0 \\ \infty & t = 0 \end{cases} \tag{A2.2}$$

such that

$$\int_{-\infty}^{\infty} \delta(t)\mathrm{d}t = 1.$$

But while this is often intuitively helpful, it is mathematically meaningless.

The Dirac delta function can also be regarded (e.g. Schwarz and Friedland, 1965) as the limit, as $\epsilon \to 0$, of a pulse of width ϵ and height $1/\epsilon$ (i.e. unit area) defined by

$$u(t) = \begin{cases} 1/\epsilon & 0 < t < \epsilon \\ 0 & \text{otherwise} \end{cases}$$

242

THE DIRAC DELTA FUNCTION

This definition is also not mathematically rigorous, but heuristically useful. In particular, control engineers can approximate such an impulse by an impulse with unit area whose duration is short compared with the least significant time constant of the response to the linear system being studied.

Even though $\delta(t)$ is a generalized function, it can often be handled as if it were an ordinary function except that we will be interested in the values of integrals involving $\delta(t)$ and never in the value of $\delta(t)$ by itself.

The derivative $\delta'(t)$ of $\delta(t)$ can also be defined by

$$\int_{-\infty}^{\infty} \delta'(t)\phi(t)\mathrm{d}t = -\phi'(0) \tag{A2.3}$$

where $\phi'(0)$ is the derivative of $\phi(t)$ evaluated at $t = 0$. The justification for A2.3 depends on integrating by parts as if $\delta'(t)$ and $\delta(t)$ were ordinary functions and using (A2.2) to give

$$\int_{-\infty}^{\infty} \delta'(t)\phi(t)\mathrm{d}t = -\int_{-\infty}^{\infty} \delta(t)\phi'(t)\mathrm{d}t$$

and then using (A2.1). Higher derivatives of $\delta(t)$ may be defined in a similar way.

The delta function has many useful properties (see, for example, Hsu, 1967).

Exercises

1. The function $\phi(t)$ is continuous at $t = t_0$. If $a < b$, show that

$$\int_{a}^{b} \delta(t - t_0)\phi(t)\mathrm{d}t = \begin{cases} \phi(t_0) & a < t_0 < b \\ 0 & t_0 < a, t_0 > b. \end{cases}$$

2. The function $\phi(t)$ is continuous at $t = 0$. Show that $\phi(t)\delta(t) = \phi(0)\delta(t)$.

APPENDIX III

Covariance

Any reader who is unfamiliar with the laws of probability, distributions, expectation, and basic statistical inference should consult a more elementary text such as Chatfield, C. (1983), *Statistics for Technology*, 3rd edition, Chapman and Hall; or Meyer, P. L. (1970), *Introductory Probability and Statistical Applications*, 2nd edition, Addison-Wesley.

The idea of covariance is particularly important in the study of time series, and will now be briefly revised.

Suppose two random variables, X and Y, have means μ_x, μ_y, respectively. Then the covariance of X and Y is defined to be

$$\text{Cov}(X,Y) = E\{(X - \mu_x)(Y - \mu_y)\}.$$

If X and Y are independent, then

$$E[(X - \mu_x)(Y - \mu_y)] = E(X - \mu_x)E(Y - \mu_y)$$
$$= 0$$

so that the covariance is zero. If X and Y are *not* independent, then the covariance may be positive or negative depending on whether 'high' values of X tend to go with 'high' or 'low' values of Y.

Covariance is a useful quantity for many mathematical purposes, but it is difficult to interpret as it depends on the units in which X and Y are measured. Thus it is often useful to standardize the covariance between two random variables by dividing by the product of their respective standard

eviations to give a quantity called the *correlation*
oefficient. It can easily be shown that the correlation
oefficient must lie between ±1.

Given N pairs of observations, $\{(x_i, y_i); i = 1 \text{ to } N\}$, the
sual unbiased estimator of the covariance between two
ariables is given by

$$s_{xy} = \sum_{i=1}^{N} (x_i - \bar{x})(y_i - \bar{y})/(N - 1).$$

he usual estimate of the correlation coefficient between the
wo variables is given by equation (2.1).

If X and Y are random variables from the same stochastic
rocess at different times, then the covariance coefficient is
alled an *auto*covariance coefficient, and the correlation
oefficient is called an *auto*correlation coefficient. If the
rocess is stationary, the standard deviations of X and Y will
e the same and their product will be the variance of X (or of
).

Some Worked Examples

In this appendix, some detailed worked examples are presented to give more of the flavour of practical time-series analysis. First I make some general remarks on the difficulties involved in time-series analysis as well as some comments on the MINITAB computer package.

General comments

I have been fascinated by Time-Series Analysis since taking a course on the subject as a postgraduate student. Like most of my fellow students, I found it difficult but feasible to understand the time-domain material on autocorrelation functions, ARMA models, etc., but found spectral analysis initially incomprehensible. Even today, despite a well-above average amount of practice, I still find time-series analysis more difficult than most other statistical techniques (except perhaps multivariate analysis). This may or may not be comforting to the reader! It is interesting to assess why difficulties arise, and this we now do.

The special feature of time-series data is that they are correlated in time rather than independent. Now most classical statistical inference requires zero correlation between the observations and it is generally true that correlated observations are more difficult to handle. The analysis of time-series data is often further complicated by the possible presence of trend and seasonal variation which

can be hard to estimate and/or remove. Other practical
problems include missing observations and outliers. But
perhaps the major problem arises from the large subjective
element involved in trying to select an appropriate time-series
model for a given set of data. The interpretation of the
correlogram and the sample spectrum is not easy, even after
much practice. If the analysis of a single (linear) time series is
not easy, then the analysis of multivariate, non-stationary
and non-linear time series must be even more difficult, and
the reader is not advised to tackle such problems until he has
had plenty of experience with standard univariate problems.

The MINITAB computer package

Until recent years, the difficulties of time-series analysis were
exacerbated by computing problems. From my point of view,
the most welcome recent advance in time-series analysis has
been the arrival of a time-series section of a computer package
called MINITAB. This easy-to-use interactive package will
perform most of the usual statistical tasks such as the
calculation of summary statistics, regression and analysis of
variance. It is ideal for the student learning statistics, as well
as being very useful for the more advanced user. In time
series, it will plot data, calculate the autocorrelation function
(ac.f.), the partial ac.f., and the cross-correlation function
between pairs of series. It will also calculate various
differences of the given series and their correlation functions.
It will also fit seasonal or non-seasonal ARIMA models and
compute forecasts. It does not (at the time of writing) carry
out spectral analysis. Further information may be obtained
by writing to: Minitab Project, Statistics Department,
215 Pond Laboratory, Pennsylvania State University,
University Park, PA 16802, USA.

In addition to MINITAB, various other packages and
algorithms are available for carrying out various forms of
time-series analysis. These include the BMDP package and the

NAG algorithms. I shall not attempt to review them here, partly to save space and partly because such information dates rapidly. If MINITAB is inadequate or unavailable, the reader is advised to check with his/her computer unit as to what is available.

Examples

I have selected three examples to try and give the reader further guidance on practical time-series analysis. In addition to the examples given below, there are several more scattered through the text including three rather brief forecasting problems in Section 5.5 and a more detailed spectral analysis example in Section 7.9.

Many other examples can be found in the literature and I note particularly that many univariate and multivariate examples of the Box–Jenkins approach may be found in Jenkins (1979).

Example 1 Monthly air temperature at Recife Table 1 shows the air temperature at Recife in Brazil, in successive months over a 10-year period. The individual observations are obtained by averaging over a one-month period.

The objective of our analysis is simply to describe and understand the variation in the data. In fact this set of data has been used throughout the book as an example and it is convenient to bring the results together here.

The first step is to plot the data, and this has been done in Fig. 1.2 on page 2. The graph shows a large regular seasonal variation with little or no trend. As one would expect, the spectrum of the raw data shows a large peak at a frequency of one cycle per year, see Fig. 7.5(a) on page 161. The frequency of one cycle per year corresponds to seasonal variation. As the seasonal variation is such a high proportion of the total variation (about 85%), the spectrum tells us very little that is not obvious in the time-plot. The correlogram

Table 1   Average monthly air temperatures at Recife for 1953—1962

Year	Jan	Feb	Mar	Apr	May	June	July	Aug	Sept	Oct	Nov	Dec
1953	26.8	27.2	27.1	26.3	25.4	23.9	23.8	23.6	25.3	25.8	26.4	26.9
1954	27.1	27.5	27.4	26.4	24.8	24.3	23.4	23.4	24.6	25.4	25.8	26.7
1955	26.9	26.3	25.7	25.7	24.8	24.0	23.4	23.5	24.8	25.6	26.2	26.5
1956	26.8	26.9	26.7	26.1	26.2	24.7	23.9	23.7	24.7	25.8	26.1	26.5
1957	26.3	27.1	26.2	25.7	25.5	24.9	24.2	24.6	25.5	25.9	26.4	26.9
1958	27.1	27.1	27.4	26.8	25.4	24.8	23.6	23.9	25.0	25.9	26.3	26.6
1959	26.8	27.1	27.4	26.4	25.5	24.7	24.3	24.4	24.8	26.2	26.3	27.0
1960	27.1	27.5	26.2	28.2	27.1	25.4	25.6	24.5	24.7	26.0	26.5	26.8
1961	26.3	26.7	26.6	25.8	25.2	25.1	23.3	23.8	25.2	25.5	26.4	26.7
1962	27.0	27.4	27.0	26.3	25.9	24.6	24.1	24.3	25.2	26.3	26.4	26.7

Fig. 2.4(a) on p. 29) also shows little apart from the seasonal variation.

In order to make any further progress we must remove the seasonal variation from the data. This can be done by seasonal differencing (see Section 4.6) or by simply calculating the overall January average, the overall February average, and so on, and then subtracting each individual value from the appropriate monthly average. We adopted the latter procedure in this case. The correlogram of the seasonally adjusted data is shown in Fig. 2.4(b) and this shows that the first three coefficients are significantly different from zero. This indicates some short-term correlation in that a month which is say warmer than average for that month is likely to be followed by one or two further months which are also warmer than their appropriate average.

The periodogram of the seasonally adjusted data is shown in Fig. 7.5(c) and exhibits wild fluctuations which are typical of periodograms. Nothing can be learnt from this graph. To estimate the underlying spectrum, we proceed by either smoothing the periodogram or by computing an appropriate weighted transform of the autocorrelation function. The latter approach was used with a Tukey window to produce Fig. 7.5(b). The spectrum shows something of a peak at zero frequency indicating the possible presence of some trend or low-frequency variation, but there is otherwise little to note in the graph.

We summarize our analysis by saying that the series contains high seasonal variation together with a limited short-term correlation effect. The analysis has been rather negative in that we have found little of interest apart from what is evident in the time-plot of the data. It is, in fact, a fairly common occurrence in statistics to find that a very simple technique provides most of the information, and this emphasizes the importance of starting any time-series analysis by drawing a time-plot.

Example 2 Yield on short-term government securities Table 2 shows the yield on British short-term government securities in successive months over a 21-year period mainly in the 1950 and 1960s. This series was previously analysed in Chatfield (1978a). The problem is to find a suitable time-series model for the data and compute forecasts for up to 12 months ahead.

The first step is, as always, to plot the data and this has been done in Fig. A.1.

The graph shows some interesting features. Firstly, we note that there is no discernible seasonal variation. Secondly, we note a marked trend in the data from about 2% at the beginning of the series to over 7% at the end of the series. However, the trend is by no means regular and it would be a gross oversimplification to simply fit a straight line to the data. Thus the use of a trend-and-seasonal model, as in the Holt—Winters forecasting procedure, or the use of linear regression seems inappropriate. With a single important time series like this, it is worth investing some effort on the analysis and the use of the Box—Jenkins procedure is indicated.

The given time series is clearly non-stationary and some sort of differencing is required. To confirm this, the autocorrelation function of the data was calculated and,

Table 2 Yield on short-term government securities

Year	I	II	III	IV	V	VI	VII	VIII	IX	X	XI	XII
1	2.22	2.23	2.22	2.20	2.09	1.97	2.03	1.98	1.94	1.79	1.74	1.86
2	1.78	1.72	1.79	1.82	1.89	1.99	1.89	1.83	1.71	1.70	1.97	2.21
3	2.36	2.41	2.92	3.15	3.26	3.51	3.48	3.16	3.01	2.97	2.88	2.91
4	3.45	3.29	3.17	3.09	3.02	2.99	2.97	2.94	2.84	2.85	2.86	2.89
5	2.93	2.93	2.87	2.82	2.63	2.33	2.22	2.15	2.28	2.28	2.06	2.54
6	2.29	2.66	3.03	3.17	3.83	3.99	4.11	4.51	4.66	4.37	4.45	4.58
7	4.58	4.76	4.89	4.65	4.51	4.65	4.52	4.52	4.57	4.65	4.74	5.10
8	5.00	4.74	4.79	4.83	4.80	4.83	4.77	4.80	5.38	6.18	6.02	5.91
9	5.66	5.42	5.06	4.70	4.73	4.64	4.62	4.48	4.43	4.33	4.32	4.30
10	4.26	4.02	4.06	4.08	4.09	4.14	4.15	4.20	4.30	4.26	4.15	4.27
11	4.69	4.72	4.92	5.10	5.20	5.56	6.08	6.13	6.09	5.99	5.58	5.59
12	5.42	5.30	5.44	5.32	5.21	5.47	5.96	6.50	6.48	6.00	5.83	5.91
13	5.98	5.91	5.64	5.49	5.43	5.33	5.22	5.03	4.74	4.55	4.68	4.53
14	4.67	4.81	4.98	5.00	4.94	4.84	4.76	4.67	4.51	4.42	4.53	4.70
15	4.75	4.90	5.06	4.99	4.96	5.03	5.22	5.47	5.45	5.48	5.57	6.33
16	6.67	6.52	6.60	6.78	6.79	6.83	6.91	6.93	6.65	6.53	6.50	6.69
17	6.58	6.42	6.79	6.82	6.76	6.88	7.22	7.41	7.27	7.03	7.09	7.18
18	6.69	6.50	6.46	6.35	6.31	6.41	6.60	6.57	6.59	6.80	7.16	7.51
19	7.52	7.40	7.48	7.42	7.53	7.75	7.80	7.63	7.51	7.49	7.64	7.92
20	8.10	8.18	8.52	8.56	9.00	9.34	9.04	9.08	9.14	8.99	8.96	8.86
21	8.79	8.62	8.29	8.05	8.00	7.89	7.48	7.31	7.42	7.51	7.71	7.99

Figure A.1   Yield on short-term government securities

as expected, the coefficients at low lags were all 'large' and positive as in Fig. 2.3. The values at lags one to five were 0.97, 0.94, 0.92, 0.89 and 0.85 while the value at lag 24 was still as high as 0.34.

The simplest type of differencing procedure is to take first differences of the data and this was done next. The autocorrelation function of the first differences turned out to be of a particularly simple form. The coefficient at lag one was 0.31, which is significantly different from zero. Here, we use the rule that values outside the range $\pm 2/\sqrt{N}$ are significantly different from zero, where N is the number of terms in the differenced series. However, all subsequent coefficients up to about lag 20 were *not* significantly different from zero.

This tells us two things. Firstly, the differenced series is now stationary and so no further differencing is required. Secondly, the fact that there is only one significant coefficient at lag one tells us that a very simple ARIMA model can be fitted. The autocorrelation function of the first-order moving average model (see Section 3.4.3) has the same form as the observed correlogram, and we can therefore try fitting on ARIMA (0,1,1) model to the observed data (or equivalently an ARMA (0,1) model to the first differences).

Using MINITAB this model was fitted giving

$$\nabla X_t = Z_t + 0.37 Z_{t-1}$$

A similar model was also tried with an added constant on the right hand side of the above equation, in view of the trend in the series. The residual sum of squares was not in fact much smaller and the estimated constant was not significantly different from zero. This is perhaps somewhat surprising at first sight but variation due to the trend is 'small' compared with the variations about the trend.

We are now in a position to compute forecasts. The above model was actually fitted to the first 17 years data with a view to comparing forecasts for the last four years with the observed values. However, the simple form of the fitted model means that the point forecasts from a given origin (e.g. the end of the 17th year) are all the same, although the confidence intervals do of course get wider as the distance ahead increases. For our fitted model, we wish to forecast $X_{N+1}, X_{N+2}, \ldots$ given data up to time N. Now $X_{N+1} = Z_{N+1} + 0.37 Z_N$. At time N, Z_{N+1} is unknown and is estimated as zero, while Z_N is estimated by the residual at time N. This cannot be written out explicitly as it involves the residual at time $(N - 1)$ and is obtained iteratively. Nevertheless, we find that our forecast of X_{N+1} (and of $X_{N+2}, X_{N+3}, \ldots$) is given by $X_N + 0.37 \times$ (residual at time N). The difference between this forecast and X_{N+1} gives the residual at time $(N + 1)$ which can in turn be used to compute forecasts from origin $(N + 1)$.

It is interesting to note that the above forecasting procedure is in fact equivalent to exponential smoothing (see Section 5.2.2), which we have already noted is optimal for an ARIMA (0,1,1) model. As exponential smoothing is a special case of the Holt–Winters forecasting procedure with an appropriate choice of starting values, it turns out that the latter procedure could have been used after all. Nevertheless, the careful choice of an ARIMA model via the Box–Jenkins procedure is much more satisfying and reliable.

This example is one of the simplest Box–Jenkins analyses that the reader is likely to come across. The only simpler class of time series are many share price series where it is often found that the first differences have a correlogram which cannot be distinguished from that of a completely random series. In other words none of the autocorrelation coefficients are significantly different from zero. This suggests that the (undifferenced) time series can be described by a random walk model.

In the next example, we present a much more difficult example of a Box–Jenkins analysis.

Example 3 Airline passenger data   Table 3 shows the monthly totals of international airline passengers from January 1949–December 1960. This famous set of data has been analysed by many people including Box and Jenkins (1970, Chapter 9). The problem is to fit a suitable model to the data and produce forecasts for up to one year ahead.

As always we begin by plotting the data as in Fig. A.2. This shows a clear seasonal pattern as well as an upward trend. The magnitude of the seasonal variation increases at the same sort of rate as the yearly mean levels and this indicates that a *multiplicative* seasonal model is appropriate (see p. 15).

It is relatively straightforward to use the Holt–Winters forecasting procedure on this data. This procedure can cope with trend and with additive or multiplicative seasonality (see Section 5.2.3 and Chatfield, 1978a). The details of the analysis will not be given here as no particular difficulties arise and reasonable forecasts result. The reader who prefers

Table 3   Monthly totals (X_t) of international airline passengers in thousands from January 1949–December 1960

Year	Jan	Feb	Mar	Apr	May	June	July	Aug	Sept	Oct	Nov	Dec
1949	112	118	132	129	121	135	148	148	136	119	104	118
1950	115	126	141	135	125	149	170	170	158	133	114	140
1951	145	150	178	163	172	178	199	199	184	162	146	166
1952	171	180	193	181	183	218	230	242	209	191	172	194
1953	196	196	236	235	229	243	264	272	237	211	180	201
1954	204	188	235	227	234	264	302	293	259	229	203	229
1955	242	233	267	269	270	315	364	347	312	274	237	278
1956	284	277	317	313	318	374	413	405	355	306	271	306
1957	315	301	356	348	355	422	465	467	404	347	305	336
1958	340	318	362	348	363	435	491	505	404	359	310	337
1959	360	342	406	396	420	472	548	559	463	407	362	405
1960	417	391	419	461	472	535	622	606	508	461	390	432

a nice simple procedure is advised to use this method for this data.

Following Box and Jenkins (1970) and other authors, we will also try to fit an ARIMA model and consider the many problems which arise.

Figure A.2   Monthly totals of international airline passengers (in thousands)

255

The first question which arises is whether or not to transform the data. The question of transformations is briefly discussed in Section 2.4, and, in order to make the seasonal effect additive, it looks as though we should take logarithms of the data. Box and Jenkins (1970, p. 303) also took logarithms of the data. To justify this, they simply say that 'logarithms are often taken before analysing sales data, because it is *percentage* fluctuation which might be expected to be comparable at different sales volumes'. This seems rather unconvincing and the long discussion of Chatfield and Prothero (1973) indicates some of the problems involved in transformation. Nevertheless, it seems fairly clear that we should take logarithms of the data in this example and so we shall proceed with the analysis on this assumption.

As an aside, we mention that the logarithmic transformation is a special case of an increasingly popular class of transformations called the Box–Cox transformation. For a given λ, the transformed value, $X_t^{(\lambda)}$, is given by

$$X_t^{(\lambda)} = \begin{cases} (X_t^{\lambda} - 1)/\lambda & \lambda \neq 0 \\ \log X_t & \lambda = 0 \end{cases}$$

This is effectively just a power transformation when $\lambda \neq 0$, since the constants are introduced to make $X_t^{(\lambda)}$ a continuous function of λ at $\lambda = 0$. It is instructive to note that Nelson and Granger (1979) found little improvement in forecast performance using the Box–Cox transformation.

Returning to the example, it is easy to plot the logged data using MINITAB which shows that the size of the seasonal variation is now roughly constant.

We now use MINITAB to plot the autocorrelation function (ac.f.) of the raw logged data and of various differenced series in order to see what form of differencing is required. The first 36 coefficients of the ac.f. of the logged data are all positive as a result of the obvious trend in the data. Some sort of differencing is clearly required. With monthly seasonal data, the obvious operators to try are, ∇, ∇_{12}, $\nabla\nabla_{12}$ and ∇_{12}^2.

SOME WORKED EXAMPLES

Letting $Y_t (= \log X_t)$ denote the logarithms of the observed data, we start by looking at the ac.f. of ∇Y_t. With N observations in the differenced series a useful rule-of-thumb for deciding if an autocorrelation coefficient is significantly different from zero is to see if its modulus exceeds $2/\sqrt{N}$. Here the critical value is 0.17 and we find significant coefficients at lags 1, 4, 8, 11, 12, 13, 16, 20, 23, 24, 25 and so on. There is no sign that the ac.f. is damping out and further differencing is required. For $\nabla_{12} Y_t$ the first nine coefficients are all positive and significantly different from zero, after which there is a long run of negative coefficients. The series is still non-stationary and so we next try $\nabla\nabla_{12} Y_t$. The first 26 coefficients are shown in Fig. A.3 in the same form as presented by the MINITAB package. (One needs to look at more than two seasonal cycles of coefficients. I actually looked at the first 38 coefficients). The number of terms in the differenced series is now $(144 - 12 - 1) = 131$ and an approximation critical value is $2/\sqrt{131} = 0.18$. We

Figure A.3   The autocorrelation function of $\nabla\nabla_{12} \log_e X_t$ as printed by MINITAB

note 'significant' values at lags 1, 3, 12 and 23, but most of the other values are 'small' and there is no evidence of non-stationarity. Thus we choose to fit an ARIMA model to $\nabla\nabla_{12} Y_t$.

It has to be admitted that our choice of the differencing operator is somewhat subjective and some authors prefer a more objective approach. One possibility is to choose the differencing operator so as to minimize the variance of the differenced series, but we did not try this here.

In order to identify a suitable ARMA model for $\nabla\nabla_{12} Y_t$, we need to calculate the partial ac.f. as well as the ac.f. (This writer actually prefers to use the inverse ac.f — see page 232 — but this is not available in MINITAB). The partial ac.f. is given in Fig. A.4. As for the ac.f., coefficients whose moduli exceed 0.18 may, as a first approximation be taken to be significantly different from zero. In the partial ac.f. we note significant values at lags 1, 3, 9 and 12. When

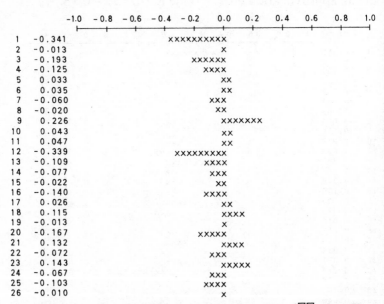

Figure A.4   The partial autocorrelation function of $\nabla\nabla_{12}\log_e X_t$ as produced by MINITAB

'significant' values occur at unusual lagged values such as
9 they are usually ignored unless there is external information
as to why such a lag should be important.

We should now be in a position to identify an appropriate
seasonal ARIMA model to fit to the data. This means that we
want to assess values of p, q, P and Q in the model defined by
equation (4.16) on p. 74. The seasonal values P and Q are
assessed by looking at the values of the ac.f. and partial ac.f.
at lags 12, 24, 36, . . . In this case the values are 'large' at lag
12 but 'small' at lags 24 and 36, indicating no autoregressive
terms but one moving average term. Thus we take $P = 0$ and
$Q = 1$. The values of the non-seasonal values p and q are
assessed by looking at the first few values of the ac.f. and
partial ac.f. The only 'significant' values are at lags 1 and 3
and these values are not easy to interpret. An AR(3) model,
as suggested by the partial ac.f., would generally have a
slowly decaying ac.f., while the MA(3) model, suggested by
the ac.f., would generally have a slowly decaying partial
ac.f. Noting that the coefficients at lag 3 are only just
'significant', we could, as a first try, take just one moving
average term and set $p = 0$ and $q = 1$. If we now work out the
standard error of the autocorrelation coefficient at lag 3
using a more exact formula, we find that it is not in fact
significantly different from zero. This more exact formula
(Box and Jenkins, 1970, p. 315) assumes that a MA(1) model
is appropriate rather than a completely random series for
which the formula $1/\sqrt{N}$ is appropriate. This result gives us
more reliance on the choice of $p = 0$ and $q = 1$. Thus we now
ask MINITAB to fit a seasonal ARIMA model with ($p = 0$,
$d = 1$, $q = 1$) and ($P = 0$, $D = 1$, $Q = 1$).

Setting $W_t = \nabla\nabla_{12} \log X_t$, the fitted model turns out to be

$$W_t = (1 - 0.396B)(1 - 0.614B^{12}) a_t$$

MINITAB readily provides estimates of the residual
variation as well as a variety of diagnostic checks. For example,
one should inspect the ac.f. of the residuals to see that 'none'
of the coefficients are significantly different from zero (but

remember that about 1 in 20 coefficients will be 'significant' at the 5% level under the null hypothesis that the residuals are random). One can also plot the observed values in the time series against the one-step-ahead forecasts of them. In this example there is no evidence that our fitted model is inadequate and so no alternative models will be tried.

MINITAB will readily provide forecasts of the logged series for up to 12 months ahead. The first three forecasts are 6.110, 6.056 and 6.178. These values need to be anti-logged to get forecasts for the original series.

I hope that this example has given some indication of the type of problem which arises in a Box—Jenkins ARIMA analysis and the sort of subjective judgement which has to be made. A critical assessment of the method has been made in Section 5.4. The method sometimes works well as in Example 2 above. However, for the airline data, the variation is dominated by trend and seasonality and yet the complicate ARMA modelling is carried out on the differenced data from which most of the trend and seasonality have removed. The reader must decide for himself if he thinks the method has been a success here. In my view, a trend-and-seasonal modelling approach (such as Holt—Winters) would not only be easier to apply but would arguably give more understandin of this particular set of data.

For the reader's interest I will now give the MINITAB commands used in the above analysis. Numbers are manipulated by MINITAB in columns labelled from $c1$ to $c50$. We read the data in $c1$, put the logs in $c3$ and the differenced series into $c6$, $c7$ and $c8$. The command 'generate 144 c2' puts integers 1 to 144 in column 2. I hope that the rest of the commands are self-explanatory. The commands are:

set c1
(read in data in rows of six observations each separated by
   one space)
generate 144 c2

```
plot c1 c2
let c3 = log (c1)
plot c3 c2
acf 38 c3
difference lag 1 c3 c6
acf 38 c6
diff 12 c3 c7
acf 38 c7
diff lag 1 c7 c8
acf 38 c8
pacf 38 c8
arima (p = 0 d = 1 q = 1) (P = 0 D = 1 Q = 1) s = 12 c3 res in
    c10 pred c11
acf 38 c10
plot c3 c11
arima (0 1 1) (0 1 1) s = 12 c3;
forecast 12 c12.
stop
```

Concluding remarks

I hope that the examples have illustrated some of the
practical difficulties which arise in time-series analysis,
particularly in regard to transformations, differencing and
the subjective interpretation of correlograms and spectra.
I also hope that the examples have demonstrated how the
choice of analysis procedure depends on the objectives
which in turn depend on the particular practical situation.
For example, if one simply wants to describe a given time
series and develop some physical understanding of it, then
a trend-and-seasonal model may be appropriate. On the
other hand, if the main objective is to produce forecasts,
then an ARIMA model may be appropriate. Clearly a
careful statement of objectives is vital.

One important point which was not discussed in the
examples is the possible use of ancillary information. In

Example 2, for instance, we have constructed a univariate model for the yield on short-term government securities which takes no account of other economic variables. As discussed in Chapter 5, it is not always possible to incorporate such ancillary information into a formal model but neither should the information be ignored. It may for example be advisable to adjust the forecasts subjectively if one knows that a change in bank rate or an oil crisis is imminent. This is just one example of how one should always be prepared to use one's common sense in carrying out a time-series analysis, or indeed any statistical investigation.

References

AKAIKE, H. (1967) *Some problems in the application of the cross-spectral method. In* Harris (1967).

AKAIKE, H. (1968) On the use of a linear model for the identification of feedback systems, *Annals of the Inst. Statist. Math.,* **20,** 425–439.

AKAIKE, H. (1973) Contribution to the discussion of Chatfield and Prothero (1973). *J. R. Statist. Soc. A,* **136,** 330.

AKAIKE, H. (1978) Time-series analysis and control through parametric models. *In Applied Time Series Analysis,* ed. D. F. Findley. New York: Academic Press.

ANDERSON, T. W. (1971) *The Statistical Analysis of Time Series.* New York: Wiley.

ARMSTRONG, J. S. and GROHMAN, M. C. (1972) A comparative study of methods for long-range forecasting, *Man. Sci.,* **19,** 211–221.

ASTROM, K. J. (1970) *Introduction to Stochastic Control Theory.* New York: Academic Press.

ASTROM, K. J. and BOHLIN, T. (1966) Numerical identification of linear dynamic systems from normal operating records, *In Theory of Self-Adaptive Control Systems,* ed. P. M. Hammond, pp. 94–111, New York: Plenum Press.

ASTROM, K. J. and EYKHOFF, P. (1971) System identification – a survey, *Automatica,* **7,** 123–162.

BARKSDALE, H. C. and GUFFEY, H. J. Jr., (1972) An illustration of cross-spectral analysis in marketing, *J. of Marketing Research,* **9,** 271–278.

BARTLETT, M. S. (1966) *Stochastic Processes,* 2nd edn. Cambridge Univ. Press.

BATES, J. M. and GRANGER, C. W. J. (1969) The combination of forecasts, *Op. Res. Quart.,* **20,** 451–468.

BENDAT, J. S. and PIERSOL, A. G. (1971) *Random Data: Analysis and Measurement Procedures.* 2nd edn. New York: Wiley.

BESAG, J. (1974) Spatial interaction and the statistical analysis of lattice systems, *J. Roy. Statist. Soc.* B, **36**, 192–236.

BHATTACHARYYA, M. N. (1974) Forecasting the demand for telephones in Australia, *Appl. Statist.*, **23**, 1–10.

BINGHAM, C., GODFREY, M. D. and TUKEY, J. W. (1967) Modern techniques of power spectrum estimation, *IEEE Trans. on Audio and Electroacoustics,* AU–15, No. 2, 56–66.

BLACKMAN, R. B. and TUKEY, J. W. (1959) *The Measurement of Power Spectra.* New York: Dover.

BLAKE, I. F. and LINDSEY, W. C. (1973) Level-crossing problems for random processes, *IEEE Trans.,* Vol. IT-19, No. 3, 295–315.

BLOOMFIELD, P. (1976) *Fourier Analysis of Time Series: An Introduction.* New York: Wiley.

BOGERT, B. P., HEALY, M. J. R. and TUKEY, J. W. (1963) The quefrency alanysis of time-series for echoes: cepstrum, pseudo-autocovariance, cross-cepstrum and saphe cracking, *In* Rosenblatt (1963).

BOX, G. E. P. and JENKINS, G. M. (1968) Some recent advances in forecasting and control. Part I, *Appl. Statist.,* **17**, 91–109.

BOX, G. E. P. and JENKINS, G. M. (1970) *Time Series Analysis, Forecasting and Control.* San Francisco: Holden-Day. (Revised edition published 1976).

BOX, G. E. P. and MACGREGOR, J. F. (1974) The analysis of closed-loop dynamic-stochastic systems, *Technometrics,* **16**, 391–398.

BOX, G. E. P. and NEWBOLD, P. (1971) Some comments on a paper of Coen, Gomme and Kendall, *J. R. Statist. Soc.* A., **134**, 229–240.

BOX, G. E. P. and PIERCE, D. A. (1970) Distribution of residual auto-correlations in autoregressive-integrated moving average time-series models, *J. Amer. Statist. Ass.,* **65**, 1509–1526.

BOX, G. E. P. and TIAO, G. C. (1975) Intervention analysis with applications to economic and environmental problems. *J. Amer. Statist. Ass.,* **70**, 70–79.

BRASS, W. (1974) Perspective in population prediction: Illustrated by the statistics of England and Wales, *J. R. Statist. Soc.,* A, **137**, 532–583.

BRILLINGER, D. R. (1965) An introduction to polyspectra, *Ann. Math. Statist.,* **36**, 1351–1374.

BRILLINGER, D. R. (1975) *Time Series: Data Analysis and Theory.* New York: Holt, Rinehart and Winston.

REFERENCES

BRILLINGER, D. R. and TIAO, G. C. (eds) (1980) *Directions in Time Series*. Institute of Mathematical Statistics.

BROWN, R. G. (1963) *Smoothing, Forecasting and Prediction.* Englewood Cliffs, N.J: Prentice-Hall.

BUTTON, K. J. (1974) A critical review of techniques used for forecasting car ownership in discrete areas, *The Statistician*, **23**, 117–128.

CHAMBERS, J. C., MULLICK, S. K. and SMITH, D. D. (1971) How to choose the right forecasting technique, *Harvard Business Review*, **49**, No. 4, 45–74.

CHAN, W. Y. T. and WALLIS, K. F. (1978) Multiple time series modelling: Another look at the mink–muskrat interaction. *Appl. Statist.*, **27**, 168–175.

CHATFIELD, C. (1974) Some comments on spectral analysis in marketing', *J. of Marketing Research*, **11**, 97–101.

CHATFIELD, C. (1977) Some recent developments in time-series analysis. *J. R. Statist. Soc.* A, **140**, 492–510.

CHATFIELD, C. (1978a) The Holt–Winters forecasting procedure. *Appl. Statist.*, **27**, 264–279.

CHATFIELD, C. (1978b) Adaptive filtering: a critical assessment. *J. Opl. Res. Soc.*, **29**, 891–896.

CHATFIELD, C. (1979) Inverse Autocorrelations. *J. R. Statist. Soc.* A. **142**, 363–377.

CHATFIELD, C. and PEPPER, M. P. G. (1971) Time-series analysis: an example from geophysical data, *Appl. Statist.*, **20**, 217–238.

CHATFIELD, C. and PROTHERO, D. L. (1973a) Box–Jenkins seasonal forecasting: problems in a case-study, *J. R. Statist Soc.* A, **136**, 295–336.

CHATFIELD, C. and PROTHERO, D. L. (1973b) A reply to some comments by Box and Jenkins, *J. R. Statist. Soc.* A, **136**, 345–352.

CLARKE, D. W. (1973) Experimental comparison of identification methods, Paper presented to UKAC 5th control convention, organized by Inst. of Measurement and Control.

CLEVELAND, W. S. and DEVLIN, S. J. (1982) Calendar effects in monthly time series: measurement and adjustment. *J. Amer. Stat. Ass.*, **77**, 520–8.

CLEVELAND, W. P. and TIAO, G. C. (1976) Decomposition of seasonal time series: A model for the Census X-11 program. *J. Amer. Statist. Ass.*, **71**, 581–587.

CLIFF, A. D. and ORD, J. K. (1975) Model building and the analysis of spatial pattern in human geography (with discussion). *J. R. Statist. Soc.* B, **37**, 297–348.

COEN, P. J., GOMME, E. J. and KENDALL, M. G. (1969) Lagged

relationships in economic forecasting, *J. R. Statist. Soc.* A, **132**, 133–163.

COOLEY, J. W., LEWIS, P. A. W. and WELCH, P. D. (1967) Historical notes on the Fast Fourier Transform, *IEEE Trans. On Aud. and Elect.*, AU-15, No. 2, 76–79.

COOLEY, J. W., LEWIS, P. A. W. and WELCH, P. D. (1969) The finite Fourier transform, *IEEE Trans. on Audio and Electroacoustics*, AU-17, No. 2, 77–85.

COUTIE, G. A. *et al.* (1964) *Short-term Forecasting, I.C.I. monograph* No. 2. Edinburgh: Oliver and Boyd.

COX, D. R. and LEWIS, P. A. W. (1966) *The Statistical Analysis of Series of Events.* London: Chapman and Hall.

COX, D. R. and MILLER, H. D. (1968) *The Theory of Stochastic Processes.* London: Chapman and Hall

COX, D. R. (1981) Statistical analysis of time series. *Scand. J. Stat.*, **8**, 93–115.

CRADDOCK, J. M. (1965) The analysis of meteorological time series for use in forecasting, *The Statistician*, **15**, 169–190.

CRAMER, H. and LEADBETTER, M. R. (1967) *Stationary and Related Stochastic Processes.* New York: Wiley.

DAVIES, N. and NEWBOLD, P. (1979) Some power studies of a portmanteau test of time series model specification. *Biometrika*, **66**, 153–155.

DURBIN, J. (1963) Trend elimination for the purpose of estimating seasonal and periodic components of time-series', *In* Rosenblatt (1963).

DURBIN, J. (1970) Testing for serial correlation in least-squares regression when some of the regressors are lagged dependent variables, *Econometrica*, **38**, 410–421.

DURBIN, J. and MURPHY, M. J. (1975) Seasonal adjustment based on a mixed additive-multiplicative model. *J. R. Statist. Soc.* A, **138**, 385–410.

DURBIN, J. and WATSON, G. S. (1950; 1951; 1971) Testing for serial correlation in least-squares regression, I, II and III, *Biometrika*, **37**, 409–428, **38**, 159–178, **58**, 1–19.

EYKHOFF, P. (1974) *System Identification.* New York: Wiley Interscience.

FULLER, A. T. (1970) *Non-linear Stochastic Control Systems.* Taylor and Francis.

FULLER, W. A. (1976) *Introduction to Statistical Time Series.* New York: Wiley.

GEARING, H. W. G. (1971) Forecasting demand, *The Statistician*, **20**,

REFERENCES

23–28.

GILCHRIST, W. (1976) *Statistical Forecasting*. London: Wiley.

GODFREY, M. D. (1965) An exploratory study of the bi-spectrum of economic time-series, *Appl. Statist.*, **14**, 48–69.

GODFREY, M. D. (1974) Computational methods for time series, *Bull. I.M.A.*, **10**, 224–227.

GOODMAN, N. R., KATZ, S., KRAMER, B. H. and KUO, M. T. (1961) Frequency response from stationary noise, *Technometrics*, **3**, 245–268.

GOTTMAN, J. M. (1981) *Time Series Analysis*. Cambridge: Cambridge Univ. Press.

GRANGER, C. W. J. (1966) The typical shape of an econometric variable, *Econometrica*, **34**, 150–161.

GRANGER, C. W. J. (1978) New classes of time series models. *The Statistician*, **27**, 237–253.

GRANGER, C. W. J. and HATANAKA, M. (1964) *Spectral Analysis of Economic Time Series*. Princeton: Princeton Univ. Press.

GRANGER, C. W. J. and HUGHES, A. O. (1968) Spectral analysis of short series – a simulation study, *J. R. Statist. Soc.* A, **131**, 83–99.

GRANGER, C. W. J. and HUGHES, A. O. (1971) A new look at some old data: the Beveridge wheat price series, *J. R. Statist. Soc.* A, **134**, 413–428.

GRANGER, C. W. J. and NEWBOLD, P. (1974) Spurious regressions in econometrics, *J. of Econometrics*, **2**, 111–120.

GRANGER, C. W J. and NEWBOLD, P. (1977) *Forecasting Economic Time Series*. New York: Academic Press.

GREGG, J. V., HOSSELL, C. V. and RICHARDSON, J. T. (1964) *Mathematical Trend Curves: An Aid to Forecasting. I.C.I. Monograph* no. 1. Edinburgh: Oliver and Boyd.

GRENANDER, U. and ROSENBLATT, M. (1957) *Statistical Analysis of Stationary Time Series*. New York: Wiley.

GRIFFITHS, J. W. R. and STOCKLIN, P. L. eds. (1973) *Signal Processing*. London: Academic Press.

GRIFFITHS, L. J. and PRIETO-DIAZ, R. (1977) Spectral analysis of natural seismic events using autoregressive techniques. *IEEE Trans. on GeoScience Electronics*, GE-15/1, 13–25.

GRIMMETT, G. R. and STIRZAKER, D. R. (1982). *Probability and Random Processes*. Oxford: Clarendon Press.

GUDMUNDSSON, G. (1971) Time-series analysis of imports, exports and other economic variables, *J. R. Statist. Soc.* A, **134**, 383–412.

GUSTAVSSON, I., LJUNG, L., and SODERSTROM, T. (1977) Identification of process in closed loop — Identifiability and accuracy aspects. *Automatica*, **13**, 59–75.

HAMON, B. V. and HANNAN, E. J. (1974) Spectral estimation of time delay for dispersive and non-dispersive systems, *Appl. Statist.*, **23**, 134–142.

HANNAN, E. J. (1960) *Time Series Analysis*. London: Methuen.

HANNAN, E. J. (1970) *Multiple Time Series*. New York: Wiley.

HARRIS, B. (ed.) (1967) *Spectral Analysis of Time Series*. New York: Wiley.

HARRIS, C. J. (1976) Problems in system identification and control. *Bull. I.M.A.*, **12**, 139–150.

HARRISON, P. J. (1965) Short-term sales forecasting, *Appl. Statist.*, **14**, 102–139.

HARRISON, P. J. and PEARCE, S. F. (1972) The use of trend curves as an aid to market forecasting, *Industrial Marketing Management*, **2**, 149–170.

HARRISON, P. J. and STEVENS, C. F. (1976) Bayesian forecasting (with discussion). *J. R. Statist. Soc.* B, **38**, 205–247.

HARVEY, A. C. (1981) *The Econometric Analysis of Time Series*. Oxford: Philip Allan.

HAUBRICH, R. A. and MACKENZIE, G. S. (1965) Earth noise, 5 to 500 millicycles per second, *J. of Geophysical Research*, **70**, No. 6, 1429–1440.

HAUGH, L. D. and BOX, G. E. P. (1977) Identification of dynamic regression (distributed lag) models connecting two time series. *J. Amer. Statist. Ass.*, **72**, 121–130.

HAUSE, J. C. (1971) Spectral analysis and the detection of lead-lag relations, *Amer. Econ. Rev.*, **61**, 213–217.

HSU, H. P. (1967) *Fourier Analysis*. New York: Simon and Schuster.

JACOBS, O. L. R. (1974) *Introduction to Control Theory*. Oxford: Clarendon Press.

JENKINS, G. M. (1979) *Practical Experiences with Modelling and Forecasting Time Series*. Jersey: Gwilym Jenkins and Partners (overseas) Ltd.

JENKINS, G. M. and WATTS, D. G. (1968) *Spectral Analysis and its Applications*. San Francisco: Holden-Day.

JONES, R. H. (1964) Spectral analysis and linear prediction of meteorological time series, *J. Applied Meteorology*, **3**, 45–52.

JONES, R. H. (1965) A re-appraisal of the periodogram in spectral analysis, *Technometrics*, **7**, 531–542.

JONES, R. H. (1974) Identification and autoregressive spectrum

REFERENCES

estimation. *Trans. IEEE on Aut. Control*, AC-19, 894–897.

JONES, R. H. (1975) Fitting autoregressions, *J. Amer. Statist. Ass.*, **70**, 590–592.

KEAY, F. (1972) *Marketing and Sales Forecasting.* London: Pergamon.

KENDALL, M. G. (1976) *Time Series.* 2nd edition London: Griffin.

KENDALL, M. G. and STUART, A. (1983) *The Advanced Theory of Statistics,* Vol. 3. 4th edn, London: Griffin.

KENNY, P. B. and DURBIN, J. (1982) Local trend estimation and seasonal adjustment of economic and social time series (with Discussion). *J. Roy. Statist. Soc. A,* **145**, 1–41.

KLEINER, B., MARTIN, R. D. and THOMSON, D. J. (1979) Robust estimation of power spectra. *J. R. Statist. Soc. B,* **41**, 313–351.

KOOPMANS, L. H. (1974) *The Spectral Analysis of Time Series.* New York: Academic Press.

LANDERS, T. F. and LACOSS, R. T. (1977) Some geophysical applications of autoregressive spectral estimates. *IEEE Trans. on GeoScience Electronics,* GE-15/1, 26–32.

LEE, Y. W. (1960) *Statistical Theory of Communication.* New York: Wiley.

LEUTHOLD, R. M., MACCORMICK, A. J. A., SCHMITZ, A. and WATTS, D. G. (1970) Forecasting daily hog prices and quantities: a study of alternative forecasting techniques. *J. Amer. Stat. Ass.,* **65**, 90–107.

LEVENBACH, H. and REUTER, B. E. (1976) Forecasting trending time series with relative growth rate models. *Technometrics,* **18**, 261–272.

LJUNG, G. M. and BOX, G. E. P. (1978) On a measure of lack of fit in time series models. *Biometrika,* **65**, 297–303.

LJUNG, G. M. and BOX, G. E. P. (1979) The likelihood function of stationary autoregressive-moving average models. *Biometrika,* **66**, 265–70.

MACKAY, D. B. (1973) A spectral analysis of the frequency of supermarket visits, *J. of Marketing Research,* **10**, 84–90.

MAKRIDAKIS, S. and WHEELWRIGHT, S. C. (1977) Adaptive filtering: an integrated autoregressive/moving average filter for time series forecasting. *Opl. Res. Q,* **28**, 425–437.

MEINHOLD, R. J. and SINGPURWALLA, N. D. (1983) Understanding the Kalman filter. *Amer. Statistician,* **32**, 123–127.

MONTGOMERY, D. C. and JOHNSON, L. A. (1977) *Forecasting and time series analysis.* New York: McGraw-Hill.

MOOERS, C. N. K. and SMITH, R. L. (1968) Continental shelf waves off Oregon, *J. of Geophysical Research,* **73**, No. 2, 549–557.

MUNK, W. H., SNODGRASS, F. E. and TUCKER, M. J. (1959) Spectra of low-frequency ocean waves, *Bull. Scripps Inst. Oceanog.*, **7**, 283–362.

MYERS, C. L. (1971) Forecasting electricity sales, *The Statistician*, **20**, 15–22.

NAYLOR, T. H., SEAKS, T. G. and WICHERN, D. W. (1972) Box–Jenkins methods: an alternative to econometric models, *Int. Statist. Rev.*, **40**, 123–137.

NEAVE, H. R. (1971) The exact error in spectrum estimators, *Ann. Math. Statist.*, **42**, 961–975.

NEAVE, H. R. (1972a) Observations on 'Spectral analysis of short series – a simulation study' by Granger and Hughes, *J. R. Statist. Soc.* A, **135**, 393–405.

NEAVE, H. R. (1972b) A comparison of lag window generators, *J. Amer. Statist. Ass.*, **67**, 152–158.

NELSON, C. R. (1973) *Applied Time Series Analysis for Managerial Forecasting.* San Francisco: Holden-Day.

NELSON, H. L. and GRANGER, C. W. J. (1979) Experience with using the Box–Cox transformation when forecasting economic time series. *J. Econometrics*, **10**, 57–69.

NERLOVE, M. and WALLIS, K. F. (1966) Use of the Durbin–Watson statistic in inappropriate situations, *Econometrica*, **34**, 235–238.

NEWBOLD, P. (1981) Some recent developments in time-series analysis. *Int. Statist. Rev.*, **49**, 53–66.

NEWBOLD, P. and GRANGER, C. W. J. (1974) Experience with forecasting univariate time-series and the combination of forecasts, *J. R. Statist. Soc.* A, **137**, 131–165.

NICHOLLS, D. F. (1977) A comparison of estimation methods for vector linear time series models. *Biometrika*, **64**, 85–90.

NOLL, A. M. (1973) The cepstrum and some close relatives, *In* Griffiths and Stocklin (1973).

OTNES, R. K. and ENOCHSON, L. (1972) *Digital Time Series Analysis.* New York: Wiley.

OTOMO, T., NAKAGAWA, T. and AKAIKE, H. (1972) Statistical approach to computer control of cement rotary kilns, *Automatica*, **8**, 35–48.

PAPOULIS, A. (1965) *Probability, Random Variables, and Stochastic Processes.* New York: McGraw-Hill.

PAGANO, M. (1972) An algorithm for fitting autoregressive schemes, *Appl. Statist.*, **21**, 274–281.

PARZEN, E. (1962) *Stochastic Processes.* San Francisco: Holden-Day.

REFERENCES

PARZEN, E. (1964) An approach to empirical time series analysis, *Radio Science J. of Research*, **68D**, 937–951.

PARZEN, E. (1980) Time series modelling, spectral analysis and forecasting. *In Directions in Time Series*, eds D. R. Brillinger and G. C. Tiao. Institute of Mathematical Statistics.

PIERCE, D. A. (1970) A duality between autoregressive and moving average processes concerning their least squares parameter estimates, *Ann. Math. Statist.*, **41**, 722–726.

PIERCE, D. A. (1977) Relationships   and the lack thereof   between economic time series, with special reference to money and interest rates. *J. Amer. Statist. Ass.*, **72**, 11–22.

PIERCE, D. A. (1980) Some recent developments in seasonal adjustment. *In Directions in Time Series*, eds D. R. Brillinger and G. C. Tiao. Institute of Mathematical Statistics.

POCOCK, S. J. (1974) Harmonic analysis applied to seasonal variations in sickness absence, *Appl. Statist.*, **23**, 103–120.

PRIESTLEY, M. B. (1971) Fitting relationships between time-series, *Bull. of the Int. Statist. Inst.*, **44**, Book 1, 295–324.

PRIESTLEY, M. B. (1981) *Spectral Analysis and Time Series*, Vols 1 and 2. London: Academic Press.

QUENOUILLE, M. H. (1957) *The Analysis of Multiple Time Series.* London: Griffin.

REID, D. J. (1975) A review of short-term projection techniques. In *Practical Aspects of Forecasting*, ed. H. A. Gordon, pp. 8–25, London: Op. Res. Soc.

RIPLEY, B. D. (1981) *Spatial Statistics.* Chichester: Wiley.

ROBERTS, J. B. and GASTER, M. (1980) On the estimation of spectra from randomly sampled signals: a method of reducing variability. *Proc. Roy. Soc.*, **A371**, 235–258.

RODEN, G. I. (1966) Low frequency sea level oscillations along the Pacific coast of N. America, *J. of Geophysical Research*, **71**, No. 20, 4755–4776.

ROSENBLATT, M. (ed) (1963) *Proceedings of Symposium on Time Series Analysis held at Brown University*, June 11–14, 1962. New York: Wiley.

SCHWARZ, R. J. and FRIEDLAND, B. (1965) *Linear Systems.* New York: McGraw-Hill.

SHIBATA, R. (1976) Selection of the order of an autoregressive model by Akaike's information criterion. *Biometrika*, **63**, 117–126.

SHISKIN, J. and PLEWES, T. J. (1978) Seasonal adjustment of the U.S. unemployment rate. *The Statistician*, **27**, 181–202.

SINGLETON, R. C. (1969) An algorithm for computing the mixed radix fast Fourier transform, *IEEE Trans. on Audio and Elect.,* Au-17, 93–103.

SNEDDON, I. H. (1972) *The Use of Integral Transforms.* New York: McGraw Hill.

SNODGRASS, F. E., GROVES, G. W., HASSELMANN, K. F., MILLER, G. R., MUNK, W. H. and POWERS, W. H. (1966) Propagation of ocean swell across the Pacific, *Phil. Trans. Roy. Soc., London,* A, **259**, 431–497.

SOLODOVNIKOV, V. V. (1965) *Statistical Dynamics of Linear Automatic Control Systems.* London: Van Nostrand.

TAYLOR, P. F. and THOMAS, M. E. (1982) Short-term forecasting: horses for courses. *J. Op. Res. Soc.,* **33**, 685–694.

TEE, L. H. and WU, S. M. (1972) An application of stochastic and dynamic models for the control of a papermaking process, *Technometrics,* **14**, 481–496.

TETLEY, H. (1946) *Actuarial Statistics,* Vol. 1. Cambridge: Cambridge Univ. Press.

TIAO, G. C. and BOX, G. E. P. (1981) Modelling multivariate time series with applications. *J. Amer. Statist. Ass.,* **76**, 802–816.

TICK, L. J. (1967) Estimation of coherency, In Harris (1967).

TOMASEK, O. (1972) Statistical forecasting of telephone time-series, *ITU Telecomm. J.,* December, 1–7.

TONG, H. (1977) Some Comments on the Canadian Lynx data. *J. R. Statist. Soc.* A, **140**, 432–436.

TUKEY, J. W. (1967) An introduction to the calculations of numerical spectrum analysis. In Harris (1967).

WAGLE, B., RAPPOPORT, J. Q. G. H. and DOWNES, V. A. (1968) A program for short-term forecasting, *The Statistican,* **18**, 141–147.

WALLIS, K. F. (1972) Testing for fourth order autocorrelation in quarterly regression equations, *Econometrica,* **40**, 617–636.

WALLIS, K. F. (1974) Seasonal adjustment and relations between variables, *J. Amer. Statist. Ass.,* **69**, 18–31.

WETHERILL, G. B. (1977) *Sampling Inspection and Quality Control.* 2nd edition, London: Chapman and Hall.

WHITTLE, P. (1953) Estimation and information in stationary time series, *Arkiv. für Mathematik,* $\dot{2}$, 423–434.

WHITTLE, P. (1963) *Prediction and Regulation.* London: E.U.P.

WIENER, N. (1949) *Extrapolation, Interpolation, and Smoothing of Stationary Time-Series.* Cambridge, Mass.: M.I.T. Press.

WINTERS, P. R. (1960) Forecasting sales by exponentially weighted moving averages, *Man. Sci.,* **6**, 324–342.

REFERENCES

WOLD, S. (1974) Spline functions in data analysis, *Technometrics,* **16,** 1–11.

YAGLOM, A. M. (1962) *An Introduction to the Theory of Stationary Random Functions.* Englewood Cliffs, N.J.: Prentice-Hall.

YOUNG, P. (1974) Recursive approaches to time series analysis, *Bull. I.M.A.,* **10,** 209–224.

Answers to Exercises

Chapter 2

1. (b) There are various ways of assessing the trend and seasonal effects. The simplest method is to calculate the four yearly averages in 1967, 1968, 1969 and 1970, and also the average sales in each of periods I, II, ..., XIII (i.e. row and column averages). The yearly averages estimate trend, and the differences between the period averages and the overall average estimate the seasonal effects. With such a small downward trend, this rather crude procedure may well be adequate. It has the advantage of being easy to understand and to compute. A more sophisticated approach is to calculate a 13-month simple moving average, moving along one period at a time. This will give trend values for each period from period 7 to period 46. Extrapolate (by eye) the trend values to the end of the series. Calculate the difference between each observation and the corresponding trend value. Hence estimate the average seasonal effects in periods I, ..., XIII.

2. (d) $r_1 = -0.549$. Equation (2.4) is ideal if computation is performed on a computer. If a calculator is used, note that

$$\sum_{t=1}^{N-1} (x_t - \bar{x})(x_{t+1} - \bar{x}) = \sum_{t=1}^{N-1} x_t x_{t+1}$$

$$- \bar{x} \left(\sum_{t=1}^{N-1} x_t + \sum_{t=2}^{N} x_t \right) + (N-1)\bar{x}^2.$$

For this particular set of data, $\bar{x}$ happens to be one, and it is easiest to subtract one from each observation.

3. $\bar{x} \to 0$ as $N \to \infty$. $r_k \to \dfrac{\sum \cos \omega t \cos(t+k)\omega}{\sum \cos^2 \omega t}$

274

ANSWERS TO EXERCISES

Use the trigonometric result:

$$2 \cos A \cos B = \cos(A + B) + \cos(A - B).$$

4. $\pm 2/\sqrt{N} = \pm 0.1$. Thus r_7 is just 'significant', but unless there is some physical explanation for an effect at lag 7, there is no real evidence of non-randomness as one expects 1 in 20 values of r_k to be 'significant' when data are random.

Chapter 3

1.
$$\rho(k) = \begin{cases} 1 & k = 0 \\ 0.56/1.53 & k = \pm 1 \\ -0.2/1.53 & k = \pm 2 \\ 0 & \text{otherwise} \end{cases}$$

3. $\text{Var}(X_t)$ is not finite. $Y_t = Z_t + (C - 1)Z_{t-1}$.

$$\rho_Y(k) = \begin{cases} 1 & k = 0 \\ (C - 1)/[1 + (C - 1)^2] & k = \pm 1 \\ 0 & \text{otherwise} \end{cases}$$

4. $\rho(k) = 0.7^{|k|}$ $k = 0, \pm 1, \pm 2, \ldots$

6. We must have $\left| \dfrac{\lambda_1 \pm \sqrt{(\lambda_1^2 + 4\lambda_2)}}{2} \right| < 1$.

If $\lambda_1^2 + 4\lambda_2 > 0$, can easily show $\lambda_1 + \lambda_2 < 1$, and $\lambda_1 - \lambda_2 > -1$.
If $\lambda_1^2 + 4\lambda_2 < 0$, roots are complex and we find $\lambda_2 > -1$.

8. $\gamma_Y(k) = 2\gamma_X(k) - \gamma_X(k + 1) - \gamma_X(k - 1)$.

$$\gamma_Y(k) = \begin{cases} 2 - 2\lambda & k = 0 \\ -\lambda^{|k|-1}(1 - \lambda)^2 & k = \pm 1, \pm 2, \ldots \end{cases}$$

9. $X_t = Z_t + 0.3\,Z_{t-1} + 0.3^2\,Z_{t-2} + \ldots$.

10. If B^* denotes the backward shift operator, then $A(B^*)\,X_t = Z_t$ and $X_t = B(B^*)\,Z_t$, so that $Z_t = A(B^*)B(B^*)\,Z_t$. Thus $A(B^*)B(B^*)$ is the unit operator. Hence $A(s)B(s) = 1$. Show coefficient of s^k in $\sigma_Z^2 B(s)B(1/s)$ is equal to $\gamma_X(k)$. Use $A(s)B(s) = 1$ to show equivalence with $\sigma_Z^2/A(s)A(1/s)$.

11. First evaluate $\gamma(k)$. Find

$\gamma(0) = \sigma_Z^2(1 + \beta^2 + 2\alpha\beta)/(1 - \alpha^2)$

$\gamma(1) = \sigma_Z^2(\alpha + \alpha\beta^2 + \alpha^2\beta + \beta)/(1 - \alpha^2)$

$\gamma(k) = \alpha\gamma(k - 1) \qquad k = 2, 3, \ldots$

Note that $E(X_t Z_t) = \sigma_Z^2$, and $E(X_t Z_{t-1}) = (\alpha + \beta)\sigma_Z^2$.

12. $E(X(t)) = \alpha/(\alpha + \beta)$ for all t. $\mathrm{Cov}(X(t), X(t + \tau))$

$\qquad = E[X(t)X(t + \tau)] - \alpha^2/(\alpha + \beta)^2$.

$E[X(t)X(t + \tau)] = \mathrm{Prob}[\text{both } X(t) \text{ and } X(t + \tau) \text{ are } 1]$

$\qquad\qquad = \mathrm{Prob}(X(t) = 1)\mathrm{Prob}(X(t + \tau) = 1 \mid X(t) = 1)$.

13. $\rho(k) = \dfrac{45}{38}\dfrac{3}{4}^{|k|} - \dfrac{7}{38}\dfrac{1}{4}^{|k|} \quad (k = 0, \pm 1, \pm 2, \ldots)$.

The AR(3) process is non-stationary as the equation
$(1 - B - cB^2 + cB^3) = 0$ has a root on the unit circle, namely $B = 1$.

14. $\mathrm{Cov}[Ye^{i\omega t}, \bar{Y}e^{-i\omega(t+\tau)}] = e^{-i\omega\tau} \mathrm{Cov}[Y, \bar{Y}]$ does not depend on t.

Chapter 4

2. $\Sigma(x_t - \bar{x})(x_{t-1} - \bar{x}) = \hat{\alpha}_1 \Sigma(x_{t-1} - \bar{x})^2 + \ldots +$

$\qquad\qquad\qquad\qquad\qquad \hat{\alpha}_p\Sigma(x_{t-p} - \bar{x})(x_{t-1} - \bar{x})$

$\cdots \cdots \cdots \cdots \cdots \cdots \cdots \cdots \cdots \cdots \cdots \cdots \cdots$

$\Sigma(x_t - \bar{x})(x_{t-p} - \bar{x}) = \hat{\alpha}_1\Sigma(x_{t-1} - \bar{x})(x_{t-p} - \bar{x}) + \ldots +$

$\qquad\qquad\qquad\qquad\qquad \hat{\alpha}_p\Sigma(x_{t-p} - \bar{x})^2$.

All summations are for $t = (p + 1)$ to N. These equations are the same as the Yule–Walker equations except that the constant divisor $\Sigma(x_t - \bar{x})^2$ is omitted and that γ_k is calculated from $(N - p)$ instead of $(N - k)$ cross-product terms.

3. Hint: see Section 3.4.4. $\rho(2) = \alpha_1\rho(1) + \alpha_2$; $\rho(1) = \alpha_1 + \alpha_2\rho(-1) = \alpha_1 + \alpha_2\rho(1)$. Solve for α_2.

4. Values outside $\pm 2/\sqrt{100} = \pm 0.2$ are 'significant'. i.e. r_1 and r_2. A second-order MA process has an ac.f. of this type.

ANSWERS TO EXERCISES

Chapter 5

5. Suppose we denote the model by

$$(1 - \alpha B^{12})W_t = (1 + \theta B)a_t \text{ or}$$
$$X_t = a_t + \theta\, a_{t-1} + (1 + \alpha)\, X_{t-12} - \alpha X_{t-24}.$$
Then $\hat{x}(N, 1) = (1 + \hat{\alpha})x_{N-11} - \hat{\alpha}x_{N-23} + \hat{\theta}\, a_N$
and $\hat{x}(N, k) = (1 + \hat{\alpha})x_{N+k-12} - \hat{\alpha}x_{N+k-24}\ (k = 2, 3, \ldots, 12).$

Chapter 6

1. (a) $f(\omega) = \sigma_Z^2/\pi(1 - 2\lambda \cos \omega + \lambda^2)$

   (b) $f^*(\omega) = (1 - \lambda^2)/\pi(1 - 2\lambda \cos \omega + \lambda^2)$

2. (a) $f(\omega) = \sigma_Z^2 [3 + 2(2 \cos \omega + \cos 2\omega)]/\pi$

   (b) $f(\omega) = \sigma_Z^2 [1.34 + 2(0.35 \cos \omega - 0.3 \cos 2\omega)]/\pi.$

3. The non-zero mean makes no difference to the acv.f., ac.f. or spectrum.

$$\gamma(k) = \begin{cases} 1.89\ \sigma_Z^2 & k = 0 \\ 1.2\ \sigma_Z^2 & k = \pm 1 \\ 0.5\ \sigma_Z^2 & k = \pm 2 \\ 0 & \text{otherwise} \end{cases}$$

$$\rho(k) = \begin{cases} 1 & k = 0 \\ 1.2/1.89 & k = \pm 1 \\ 0.5/1.89 & k = \pm 2 \\ 0 & \text{otherwise} \end{cases}$$

4.
$$\rho(k) = \int_0^\pi f^*(\omega) \cos \omega k\ d\omega$$

$$= -\frac{2 \cos \omega k}{\pi^2 k^2} \Big]_0^\pi$$

5. Clearly $E(X(t)) = 0$

$E[X(t)X(t + u)] = \text{Prob}[X(t) \text{ and } X(t + u) \text{ have same sign}]$
$\qquad\qquad\qquad - \text{Prob}[X(t) \text{ and } X(t + u) \text{ have opposite sign}]$

Hint: Prob(even number of changes in time u)

$$= e^{-\lambda u} \left[1 + \frac{(\lambda u)^2}{2!} + \frac{(\lambda u)^4}{4!} + \ldots \right]$$

$$= e^{-\lambda u}(e^{\lambda u} + e^{-\lambda u})/2.$$

Chapter 7

6. $\quad 2N \bigg/ \sum_{k=-M}^{+M} \tfrac{1}{4}\left(1 + \cos \frac{\pi k}{M}\right)^2 = 8N/(3M + 2) \to 8N/3M$ as $M \to \infty$.

Chapter 8

1. $\quad f_{xy}(\omega) = \dfrac{\sigma_Z^2}{\pi} \left[\beta_{11}\beta_{21} + \beta_{12}\beta_{22} + \beta_{21}e^{-i\omega} + \beta_{12}e^{+i\omega} \right].$

2. Use the fact that $\text{Var}[\lambda_1 X(t) + \lambda_2 Y(t + \tau)] \geqslant 0$ for any constants λ_1, λ_2.

$$\gamma_{xy}(k) = \begin{cases} 0.84\ \sigma_Z^2 & k = 0 \\ -0.4\ \sigma_Z^2 & k = 1 \\ +0.4\ \sigma_Z^2 & k = -1 \\ 0 & \text{otherwise} \end{cases}$$

$$c(\omega) = 0.84\ \sigma_Z^2 / \pi$$

$$q(\omega) = -0.8\ \sigma_Z^2 \sin \omega / \pi$$

$$\alpha_{xy}(\omega) = \frac{\sigma_Z^2}{\pi} \sqrt{(0.84^2 + 0.8^2 \sin^2 \omega)}$$

$$\tan \phi_{xy}(\omega) = 0.8 \sin \omega / 0.84$$

$$C(\omega) = 1$$

Chapter 9

1. Selected answers only.

(a) $\quad h_k = \begin{cases} \tfrac{1}{2} & k = \pm 1 \\ 1 & k = 0 \\ 0 & \text{otherwise} \end{cases}$

278

$$S_t = \begin{cases} 0 & t < -1 \\ \tfrac{1}{2} & t = -1 \\ 1\tfrac{1}{2} & t = 0 \\ 2 & t \geqslant 1 \end{cases}$$

$H(\omega) = \tfrac{1}{2}e^{-i\omega} + 1 + \tfrac{1}{2}e^{i\omega} = 1 + \cos \omega$

$G(\omega) = H(\omega)$ since $H(\omega)$ is real. $\phi(\omega) = 0$.

(b) $H(\omega) = \dfrac{1}{5} + \dfrac{2}{5}\cos \omega + \dfrac{2}{5}\cos 2\omega$

(c) $$h_k = \begin{cases} +1 & k = 0 \\ -1 & k = +1 \\ 0 & \text{otherwise} \end{cases}$$

(a) and (b) are low-pass filters. (c) and (d) are high-pass filters.
The combined filter has

$$H(\omega) = (1 + \cos \omega)\left(\frac{1}{5} + \frac{2}{5}\cos \omega + \frac{2}{5}\cos 2\omega\right).$$

2. (a) $H(\omega) = ge^{-i\omega\tau}$ $\qquad\qquad$ $(\omega > 0)$

   (b) $H(\omega) = g(1 - i\omega T)/(1 + \omega^2 T^2)$ $\qquad$ $(\omega > 0)$.

4. $$\gamma(k) = \begin{cases} \alpha^{|k/2|}\,\sigma_Z^2/(1 - \alpha^2) & k \text{ even} \\ 0 & k \text{ odd}. \end{cases}$$

   $f_X(\omega) = f_Z(\omega)/(1 - 2\alpha \cos 2\omega + \alpha^2).$ $\qquad$ $(0 < \omega < \pi)$.

Indexes

AUTHOR INDEX

280

INDEX

INDEX

SUBJECT INDEX

INDEX

INDEX

INDEX